# Optimization of Ozone Disinfection Systems With Fluorescent-Dyed Microspheres

*The mission of the Awwa Research Foundation (AwwaRF) is to advance the science of water to improve the quality of life. Funded primarily through annual subscription payments from over 1,000 utilities, consulting firms, and manufacturers in North America and abroad, AwwaRF sponsors research on all aspects of drinking water, including supply and resources, treatment, monitoring and analysis, distribution, management, and health effects.*

*From its headquarters in Denver, Colorado, the AwwaRF staff directs and supports the efforts of over 700 volunteers, who are the heart of the research program. These volunteers, serving on various boards and committees, use their expertise to select and monitor research studies to benefit the entire drinking water community.*

*Research findings are disseminated through a number of technology transfer activities, including research reports, conferences, videotape summaries, and periodicals.*

# Optimization of Ozone Disinfection Systems With Fluorescent-Dyed Microspheres

Prepared by:
**George Tang** and **Benito Mariñas**
University of Illinois at Urbana-Champaign, Urbana, Illinois 61801

**Susan Teefy**
(formerly with Alameda County Water District)
Water Quality and Treatment Solutions, Chatsworth, California 91311

and

**Richard Sakaji**
State of California Department of Health Services, Berkeley, California 94704-1011

Sponsored by:
**Awwa Research Foundation**
6666 West Quincy Avenue, Denver, CO 80235-3098

Published by:

Awwa Research Foundation
Advancing the Science of Water®

IWA Publishing

## DISCLAIMER

This study was funded by the Awwa Research Foundation (AwwaRF). AwwaRF assumes no responsibility for the content of the research study reported in this publication or for the opinions or statements of fact expressed in the report. The mention of trade names for commercial products does not represent or imply the approval or endorsement of AwwaRF. This report is presented solely for informational purposes.

**Published by IWA Publishing, Alliance House, 12 Caxton Street, London SW1H 0QS, UK**

Telephone: +44 (0) 20 7654 5500; Fax: +44 (0) 20 7654 5555; Email: publications@iwap.co.uk;
Web: **www.iwapublishing.com**

AwwaRF report number 90990F. Originally published by AwwaRF for its subscribers in 2003.
IWA Publishing version published 2004.



ISBN 1 84339 8850

## CONTENTS

## LIST OF TABLES

## LIST OF FIGURES

## FOREWORD

The Awwa Research Foundation is a nonprofit corporation that is dedicated to the implementation of a research effort to help utilities respond to regulatory requirements and traditional high-priority concerns of the industry. The research agenda is developed through a process of consultation with subscribers and drinking water professionals. Under the umbrella of a Strategic Research Plan, the Research Advisory Council prioritizes the suggested projects based upon current and future needs, applicability, and past work; the recommendations are forwarded to the Board of Trustees for final selection. The foundation also sponsors research projects through the unsolicited proposal process; the Collaborative Research, Research Applications, and Tailored Collaboration programs; and various joint research efforts with organizations such as the U.S. Environmental Protection Agency, the U.S. Bureau of Reclamation, and the Association of California Water Agencies.

This publication is a result of one of these sponsored studies, and it is hoped that its findings will be applied in communities throughout the world. The following report serves not only as a means of communicating the results of the water industry's centralized research program but also as a tool to enlist the further support of the nonmember utilities and individuals.

Projects are managed closely from their inception to the final report by the foundation's staff and large cadre of volunteers who willingly contribute their time and expertise. The foundation serves a planning and management function and awards contracts to other institutions such as water utilities, universities, and engineering firms. The funding for this research effort comes primarily from the Subscription Program, through which water utilities subscribe to the research program and make an annual payment proportionate to the volume of water they deliver and consultants and manufacturers subscribe based on their annual billings. The program offers a cost-effective and fair method for funding research in the public interest.

A broad spectrum of water supply issues is addressed by the foundation's research agenda: resources, treatment and operations, distribution and storage, water quality and analysis, toxicology, economics, and management. The ultimate purpose of the coordinated effort is to assist water suppliers to provide the highest possible quality of water economically and reliably. The true benefits are realized when the results are implemented at the utility level. The foundation's trustees are pleased to offer this publication as a contribution toward that end.

Optimizing the operating conditions of ozone contactors is important to provide adequate *Cryptosporidium parvum* inactivation while maintaining a low level of bromate formation. The results obtained from this research should provide an effective tool to assess inactivation efficiency, and, with this information, strategies can be developed to optimize ozone contactors.

Edmund G. Archuleta, P.E.
Chair, Board of Trustees
Awwa Research Foundation

James F. Manwaring, P.E.
Executive Director
Awwa Research Foundation

## ACKNOWLEDGMENTS

The authors would like to recognize the Alameda County Water District (ACWD), and the state of California Department of Health Services (CA DHS) for their participation in this study. This project would not have been possible without the cooperation and support of them.

The authors would like to acknowledge Lauren Luce (Polysciences Inc., Warrington, PA) and Katherine Turner (Bangslabs Inc., Fisher, IN) for their technical assistance and invaluable advice on the development and application of microspheres. Jae-Hong Kim (Georgia Institute of Technology, Atlanta, GA) is gratefully acknowledged for his efforts on mathematical modeling.

The authors also gratefully acknowledge the support and assistance of Misha Hasan, AwwaRF Project Officer, and the members of the AwwaRF Project Advisory Committee (PAC): George di Giovanni (Texas A & M University, El Paso, TX), Giovanni Widmer (Tufts University, Grafton, MA); Issam Najm, (Water Quality & Treatment Solutions Inc., Chatsworth, CA) and Lon Couillard (Milwaukee Water Works, Milwaukee, WI).

## EXECUTIVE SUMMARY

In 2001, the Environmental Protection Agency (EPA) released a draft proposal of the Long Term 2 Enhanced Surface Water Treatment Rule (LT2ESWTR). A major objective of this rule is to provide adequate control for *Cryptosporidium parvum* oocysts, a protozoan parasite that has shown great resistance to conventional disinfectants such as free chlorine and monochloramine (Drideger *et al.*, 2000; Rennecker *et al.*, 2000). Under the planned LT2ESWTR, water treatment plants will be required to monitor the occurrence of *C. parvum* oocysts in the raw water, and depending on the concentration of oocysts, up to 5.5-log of total removal/inactivation may be required.

Ozone, among other chemical disinfectants, has been demonstrated as an effective disinfectant of *C. parvum* oocysts in both laboratory-scale (Rennecker *et al.*, 1999; Li *et al.*, 2000) and pilot-scale studies (Oppenheimer *et al.*, 2000; Kim *et al.*, 2002a, b). Nevertheless, it is practically impossible to quantify the disinfection efficiency of a full-scale ozone contactor. Because the concentration of oocysts present in natural and treated waters tends to be relatively low, the dual task of recovery and viability assessment is difficult. Instead, the *CT* (the product of average dissolved ozone concentration and characteristic contact time) approach, as specified by the SWTR (USEPA, 1991), has been adopted by many water utilities to demonstrate the disinfection efficiency. However, this approach may have serious limitations because possible hydrodynamic non-idealities within the system, such as dead water zones and associated short-circuiting, might not been taken into account. Such complex hydrodynamics might cause a significant difference between actual disinfection efficiency of a full-scale system and that of an ideal system.

The formation of bromate ($BrO_3^-$) is also of particular concern when using ozone as a disinfectant. Bromate is regulated under the Stage 1 Disinfectants and Disinfection By-Products (D/DBP) Rule (USEPA, 1998a) with a maximum contaminant level (MCL) and an MCL Goal (MCLG) of 10 and 0 μg/L, respectively. As a result, the design and operation of ozone contactor systems have to achieve two seemingly countering goals: to provide adequate inactivation of *C. parvum* oocysts, and to minimize the formation of bromate.

The use of fluorescence-dyed microspheres for the assessment of disinfection efficiency in full-scale ozone contactors was first proposed by Chiou *et al.* (1997) and further developed by Mariñas *et al.* (1999). The principle of the technique is based on the idea that once the microspheres are dosed into the contactor, they would be exposed to the same hydrodynamic conditions and ozone exposure that would be experienced by oocysts. The fluorescence intensity of the microspheres would decrease due to the reaction between ozone and fluorescence dye, and this drop of fluorescence can be correlated to the inactivation kinetics of *C. parvum* oocysts.

The main objective of this study was the development and demonstration of the use of polystyrene microspheres as tools for optimizing the performance of full-scale ozone disinfection systems in terms of *C. parvum* inactivation and bromate formation minimization. More specific tasks included:

- Optimization of microspheres with a laboratory-scale semi-batch apparatus.
- Development of microspheres using a fluorescence dye certified by the National Sanitation Foundation for performing tracer test studies in drinking water.
- Development of correlation between *C. parvum* oocyst and surrogate indicator "survival" with a laboratory-scale semi-batch apparatus.
- Demonstration of surrogate method with a full-scale ozone contactor.
- Application of a mathematical model for the prediction and simulation of disinfection efficiency and bromate formation in a full-scale ozone contactor.

## DEVELOPMENT OF RHODAMINE WT MICROSPHERES

The development of a microsphere with a dye that would be more acceptable for use by the drinking water industry was an integral part of the project. Rhodamine WT was specifically chosen as a fluorescence dye because it has been certified by the National Sanitation Foundation (ANSI/NSF, 2000), and reviewed by the EPA (USEPA, 1998b) for use in drinking water. One major challenge associated to the use of Rhodamine WT was its relatively high solubility in water, and corresponding tendency to leach out from the polystyrene matrix of the particle. This problem was solved by binding the carboxylic group of the dye molecule to the amine-group on the surface of modified polystyrene microspheres. However, the resulting fluorescence intensity of this first generation of Rhodamine WT microspheres was similar to that of the background fluorescence in the raw water tested. Such impasse could be avoided by increasing the mass of Rhodamine WT by allowing the dye to penetrate deeper into the microsphere prior to cross-linking. Unfortunately, the limited resources available for this project did not allow the undertaking of such second phase of Rhodamine WT microsphere development.

## EXPERIMENTAL STUDIES WITH YG MICROSPHERES

Fluoresbrite$^{TM}$ Plain YG microspheres (Polysciences, Inc., Warrington, PA) were used in both Phase I and Phase II of the full-scale demonstration experiments. The microsphere was polystyrene based and was dyed with YG, a proprietary chemical. Preliminary experiments have shown that the initial decay of fluorescence intensity was too fast for correlation with the inactivation kinetics of *C. parvum* oocysts by the threshold concept (Chiou *et al.*, 1997). Trial-and error efforts revealed that the optimal correlation could be achieved by reducing the fluorescence intensity by about 95%. Consequently, the microsphere stock was first optimized by preozonation, which reduced the fluorescence intensity to about 5% of the original value. Then, fluorescence decay kinetics of preozonated microspheres were obtained by performing semi-batch experiments. The fluorescence intensity of the microspheres was quantified by flow cytometry, a technique capable of measuring the fluorescence intensity of individual microspheres. A correlation between the fluorescence decay and *C. parvum* inactivation was developed by the threshold concept (Chiou *et al.*, 1997), where microspheres above and below a specific fluorescence threshold are considered viable and non-viable oocysts, respectively. Such correlations were used to assess the level of *C. parvum* oocyst inactivation achieved during full-scale experiments. Batch ozonation experiments were also performed to determine the parameters for ozone decomposition kinetics for both Phase I and Phase II raw waters.

Mathematical model predictions of ozone concentration and *C. parvum* inactivation profiles through the full-scale contactor were significantly higher than the Phase I experimental data. The suspected causes for such discrepancy were changes in water quality parameters during sample storage, and clumping of microspheres. In Phase II, satisfactory matches between the mathematical model predictions and experimental data were achieved.

Batch ozonation experiments revealed the occurrence of a linear dependence of bromate formation on *CT*. The mathematical model prediction of ozone concentration, *C. parvum* inactivation, and bromate formation matched the experimental data reasonably well.

Mathematical model simulations reveal that the formation of bromate could be lowered by contactor design and configuration changes that would result in a decrease in the level of back-mixing taking place in contactor chambers. Potential contactor design modifications aimed at minimizing dispersion include changing the configuration of the diffusers, and adding vertical baffles.

## RECOMMENDATIONS TO THE WATER INDUSTRY

This study has shown that microspheres can be used as a surrogate for assessing the inactivation efficiency of an ozone contactor. It is therefore recommended that water utilities using ozone as disinfectant consider performing such test. Utilities need to check with their regulatory agencies to determine if using the microspheres for such an application would be acceptable before proceeding with the study.

This study has also shown that mathematical modeling could be used to predict the performance the contactor, if the ozone decay parameter in the specific raw water is available. Mathematical simulation revealed that the hydrodynamics of the contactor has a significant effect on *C. parvum* inactivation, but has virtually no effect on bromate formation. A PFR-like contactor would provide more inactivation for *C. parvum* inactivation than the CSTR-like contactor, while the two types of behavior would cause the same bromate formation. It is therefore recommended to modify the contactor to minimize forward dispersion or in the direction of flow which can be achieved in different ways: by modifying the configuration of the diffuser, by using a diffuser rod that would produce smaller gas bubbles, or by adding vertical baffles in the contactor.

# CHAPTER 1. INTRODUCTION

## 1.1 BACKGROUND

In 2001, the Environmental Protection Agency (EPA) released a draft proposal of the Long Term 2 Enhanced Surface Water Treatment Rule (LT2ESWTR). One of the main focuses of this new rule is the control of *Cryptosporidium parvum* oocysts, a protozoan parasite that has shown strong resistance to conventional disinfectants such as free chlorine and monochloramine (Drideger *et al.*, 2000; Rennecker *et al.*, 2000). This parasite is responsible for cryptosporidiosis, a gastro-intestinal disease that can be prolonged and even be life-threatening for immunocompromised individuals. Numerous outbreaks of this disease have been reported in various regions of the United States (Solo-Gabriele *et al.*, 1996), with the most significant outbreak occurring in 1993 in Milwaukee, Wisconsin where more than 400,000 people were infected (Kramer *et al.*, 1996). Under the new rule, water treatment plants are required to monitor the occurrence of *C. parvum* oocysts in the raw water, and depending on the concentration of oocysts, up to 5.5-log of total removal/inactivation through the water treatment plant may be required.

Among chemical disinfectants, ozone has been demonstrated as an effective disinfectant of *C. parvum* oocysts in both laboratory-scale tests (Rennecker *et al.*, 1999; Li *et al.*, 2000) and pilot-scale studies (Oppenheimer *et al.*, 2000; Kim *et al.*, 2002a, b). However, a direct quantification of the disinfection efficiency achieved in full-scale ozone contactors is practically impossible at the present time. This is due to the low concentration of oocysts occurring in raw and treated water, which makes assessing oocyst viability with current techniques not possible. Instead, the $C \times T$ (i.e., product of average dissolved ozone concentration and characteristic contact time) approach, as specified by the SWTR (USEPA, 1991), is adopted by many water utilities to demonstrate the disinfection efficiency achieved in full-scale ozone contactors. Correlations between $C \times T$ values and log inactivation levels have been proposed by the LT2ESWTR, as shown in Table 1.1 (Scharfenaker, 2002). However, this approach may have serious limitations because possible hydrodynamic non-idealities within the system, such as dead-water zones and short-circuiting, are not taken into account. Such complex hydrodynamics might cause significant discrepancy between actual disinfection efficiency of a full-scale system and that of an ideal system, even after taking into account the effect of better quantifiable hydrodynamic non-idealities such as backmixing. As a result, the use of relatively high safety factors would be needed to compensate for such uncertainties, which might in turn result in disinfection system over-design, excessive treatment cost, and greater formation of disinfection by-products.

The formation of disinfection by-products (DBPs) is an important factor when considering the use of ozone as disinfectant. Bromate ($BrO_3^-$) is perhaps of most concern. It is a reaction product formed during the ozonation of bromide-containing waters. The Stage 1 Disinfectants and Disinfection By-Products (D/DBP) Rule (USEPA, 1998a) has specified a maximum contaminant level (MCL) of 10 μg/L, and an MCL Goal (MCLG) of 0 μg/L for bromate, a known animal carcinogen. As a result, the design and operation of ozone contactor systems have to achieve the two potentially conflicting goals of providing adequate inactivation of *C. parvum* oocysts as specified by the LT2ESWTR, while maintaining bromate level below the MCL, as specified by the Stage 1 D/DBP rule.

**Table 1.1: $C \times T$ values (mg-min/L) for *Cryptosporidium parvum* inactivation by ozone as a function of water temperature as noted in the draft pre-proposal of LT2ESWTR (Scharfenaker, 2002)**

| | Based on the Geometric Mean of *C* in a Contact Chamber | | | | | | | | |
|---|---|---|---|---|---|---|---|---|---|
| Log | Water Temperature - °C | | | | | | | | |
| Inactivation | 1 | 2 | 3 | 5 | 7 | 10 | 15 | 20 | 25 |
| 0.5 | 17.4 | 15.9 | 14.4 | 11.9 | 9.8 | 7.4 | 4.6 | 2.9 | 1.8 |
| 1.0 | 34.9 | 31.7 | 28.8 | 23.8 | 19.7 | 14.8 | 9.2 | 5.7 | 3.5 |
| 1.5 | 52.3 | 47.6 | 43.3 | 35.7 | 29.5 | 22.2 | 13.8 | 8.6 | 5.3 |
| 2.0 | 69.8 | 63.4 | 57.7 | 47.4 | 39.4 | 29.6 | 18.4 | 11.4 | 7.1 |
| 2.5 | 87.2 | 79.3 | 72.1 | 59.6 | 49.2 | 37.0 | 23.0 | 14.3 | 8.9 |
| 3.0 | 104.7 | 95.2 | 86.5 | 71.5 | 59.1 | 44.4 | 27.6 | 17.1 | 10.6 |
| | Based on Measuring Only the Final Concentration in a Contact Chamber | | | | | | | | |
| Log | Water Temperature - °C | | | | | | | | |
| Inactivation | 1 | 2 | 3 | 5 | 7 | 10 | 15 | 20 | 25 |
| 0.5 | 12.8 | 11.6 | 10.6 | 8.7 | 7.2 | 5.4 | 3.4 | 2.1 | 1.3 |
| 1.0 | 25.6 | 23.3 | 21.2 | 17.5 | 14.5 | 10.9 | 6.7 | 4.2 | 2.6 |
| 1.5 | 38.4 | 34.9 | 31.8 | 26.2 | 21.7 | 16.3 | 10.1 | 6.3 | 3.9 |
| 2.0 | 51.2 | 46.6 | 42.3 | 35.0 | 28.9 | 21.7 | 13.5 | 8.4 | 5.2 |
| 2.5 | 64.0 | 58.2 | 52.9 | 43.7 | 36.2 | 27.2 | 16.9 | 10.5 | 6.5 |
| 3.0 | 76.9 | 69.9 | 63.5 | 52.5 | 43.4 | 32.6 | 20.2 | 12.6 | 7.8 |

## 1.2 MICROSPHERES AS A SURROGATE FOR *C. PARVUM* OOCYSTS

The use of fluorescence-dyed microspheres for the assessment of disinfection efficiency in full-scale ozone contactors was first proposed by Chiou *et al.* (1997) and further developed by Mariñas *et al.* (1999). A scanning electron micrograph showing a *C. parvum* oocyst and one of the two microspheres investigated in this study (Figure 1.1) shows that oocysts and microspheres have similar shapes, and their sizes are of the same order of magnitude. Although the composition of the two particles are quite different, an analogy could be established in terms of the general mass transfer and reaction steps that take place when microspheres and oocysts are exposed to dissolved ozone. Dissolved ozone transfers from bulk solution to the surface of either particle, then diffuses into and reacts with various constituents of microspheres and oocysts. The principle of the technique is based on the idea that microspheres and oocysts entering a contactor would be exposed to the same hydrodynamic conditions, water quality characteristics, and ozone exposure. The fluorescence intensity of the microspheres would decrease due to the reaction between ozone and fluorescence dye, and this drop in fluorescence could be correlated to the inactivation kinetics of *C. parvum* oocysts.

Another advantage of using microspheres as a surrogate over the $C \times T$ approach is simplicity. The $C \times T$ approach involves the performance of a tracer test to characterize the contact time $T$ ($T_{10}$ from a tracer curve) and the relatively extensive measurement of dissolved ozone concentration inside the contactor to characterize the disinfectant concentration profile. The application of microspheres as non-biological surrogates involves less complicated procedures. A test involves dosing the contactor with a step-input of microspheres suspension. Once steady-state conditions are reached, microspheres can be retrieved from the contactor effluent. The fluorescence intensity can then be analyzed, and its drop correlated to the inactivation of oocysts.

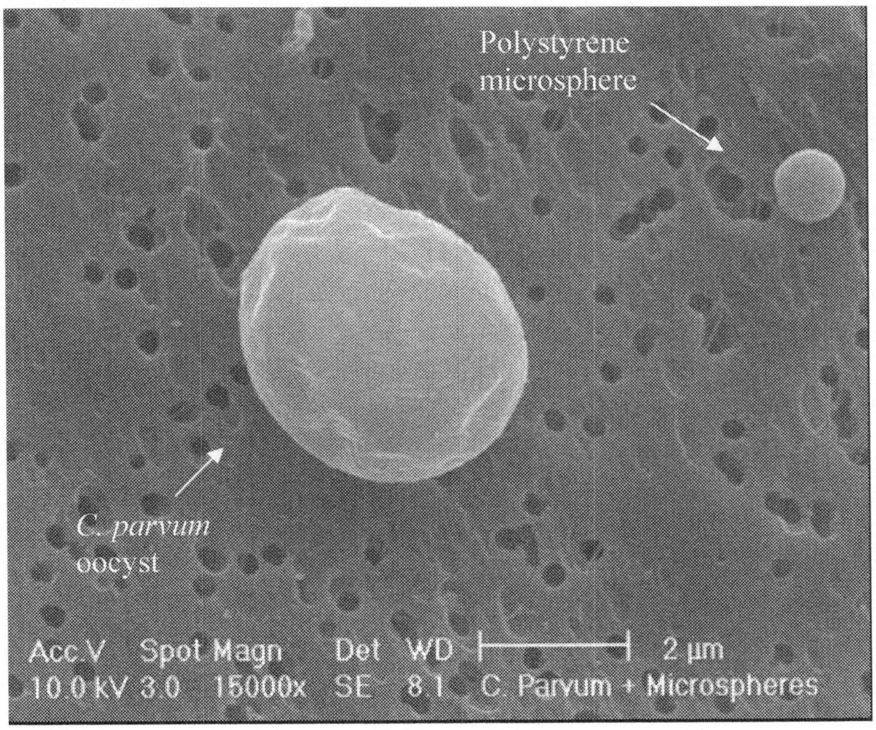


**Figure 1.1: Scanning electron micrograph of *C. parvum* oocyst and 1-µm YG microsphere**

## 1.3 RESEARCH OBJECTIVES

The main objective of this project was the development and demonstration of the use of polystyrene microspheres for assessing and optimizing the performance of full-scale ozone disinfection systems. The unit tested for the full-scale demonstration phases of this project was one of two parallel bubble-diffuser ozone contactors of the Alameda County Water District (ACWD) treatment plant located in Fremont, California. The method was based on the use of fluorescent-dyed polystyrene microspheres as non-biological surrogate for *C. parvum* oocysts. More specific tasks included:

- Optimization of surrogate with a laboratory-scale semi-batch apparatus.

- Development of microspheres using a fluorescence dye approved by the National Sanitation Foundation for water tracer test studies.

- Development of correlation between *C. parvum* oocyst and microsphere “survival" ratios with a laboratory-scale semi-batch apparatus.

- Development of recommendations for adequate protection of public health when using microspheres in full-scale treatment plants.

- Demonstration of surrogate indicator method with ACWD full-scale ozone contactor.

- Application of a mathematical model for the prediction of disinfection efficiency and bromate formation within the ACWD ozone contactor based on the results obtained from the full-scale test.

The experimental tasks described subsequently were designed for the simultaneous assessment of *C. parvum* oocyst inactivation efficiency and bromate formation in a full-scale ozone contactor. Bromate was selected because adequate control of this DBP is currently considered a requirement for the application of ozone to raw waters containing bromide. The bromide concentration in the raw water of the ACWD plant ranges seasonally from about 30 to 500 μg/L. This water was selected because of its seasonal bromide content variability, "challenging" quality, and seasonal occurrence of algae that can potentially interfere with surrogate indicator analyses.

# CHAPTER 2. MATERIALS AND METHODS

## 2.1 MICROSPHERES

### 2.1.1 Rhodamine WT microspheres

An initial attempt was made to develop a surrogate microsphere manufactured with the fluorescent dye Rhodamine WT certified for use in drinking water treatment (ANSI/NSF, 2000). Rhodamine WT microspheres were developed in collaboration with Bangs Laboratories (Fisher, IN). The carboxyl groups of the Rhodamine WT dye molecules were attached by covalent bonding to polystyrene microspheres modified with amino groups. The covalent bonding was necessary because Rhodamine WT, having relatively high solubility in water, would have otherwise leached out of the polystyrene matrix. The resulting Rhodamine WT surface-labeled microspheres had an average diameter of 0.87 micrometers. In preliminary experiments, six batches were tested, and were referred as stocks #1 to #6. The main difference between stocks was the mass of Rhodamine WT dye used during the manufacturing process. Stock #1 was manufactured with the highest mass of dye and subsequent stocks were prepared with a mass of dye of 1/100$^{th}$ that used in the preceding stock. Accordingly, stock #2 was prepared with a mass of dye that was 1/100$^{th}$ of that used for stock #1, stock #3 with 1/100$^{th}$ of that for stock #2, and so on. An additional stock of unlabeled microspheres was also tested as a blank. Table 2.1 shows the relative mass of dye used in the manufacturing of each stock. The absolute mass values were not provided because the information was considered proprietary. The stock of blank microspheres was used to correct for the background fluorescence intensity of the polystyrene matrix.

As described subsequently, the preliminary trial revealed that stock # 3 was the best candidate as a surrogate for *C. parvum* oocysts. Consequently, a request was made to the manufacturer for additional stocks with a narrower range in mass of dye comparable to stock #3 with the goal of undergoing a more refine search form the optimum surrogate. The new stocks, according to the manufacturer, were manufactured with the same procedure as the previous samples. They were referred to as samples 1/50X, 1/10X, 1X, 2X, 10X, and 50X, which corresponded to the amount of Rhodamine WT dye used relative to stock #3 for the manufacturing process. For example, 1/50$^{th}$ of the dye used for stock #3 was used to manufacture the sample 1/50X. Before the experiments, each microsphere stock was cleaned by centrifugation twice to remove the surfactant that would exert an ozone demand.

**Table 2.1: Relative Mass of Rhodamine WT Used Initially for Different Microsphere Lots**

| Stock # | Mass of dye relative to that used for stock #3 |
|---|---|
| 1 | $10^{4}$X |
| 2 | $10^{2}$X |
| 3 | 1X |
| 4 | $10^{-2}$X |
| 5 | $10^{-4}$X |
| 6 | $10^{-6}$X |

### 2.1.2 Fluoresbrite™ Plain YG Microspheres

A second non-biological surrogate chosen for this study was the Fluoresbrite™ Plain YG microspheres (Cat. # 17154, Polysciences, Warrington, PA). The microspheres were polystyrene-based. This polymer was selected because of its transparency to visible light, a crucial property for the flow cytometry analysis described subsequently. The dye, a proprietary chemical referred to as YG by the manufacturer, was hydrophobic and matched the fluorescence filter settings for fluorescein isothiocyanate (FITC), i.e., maximum excitation and emission wavelengths of 458 and 540 nm, respectively. Two different batches of microspheres were used for the two separate phases (Phases I and II) of full-scale testing described in Section 2.3. Batch I microspheres, with an average diameter of 0.93 micrometers, were used for Phase I, and Batch II microspheres, with an average diameter of 0.94 micrometers, were used for Phase II. The microsphere density of both batches was 1.045 g/mL.

## 2.2 BENCH-SCALE EXPERIMENTS

### 2.2.1 Semi-batch Experimental Setup

The semi-batch reactor system described by Rennecker *et al.* (1999) was used both in ozonation experiments for both types of microspheres tested as well as for the preozonation of YG microspheres described in Section 2.2.2. The reactor consisted of a 250-ml glass cylinder immersed in a water bath. The concentration of dissolved ozone was maintained constant by applying ozonated gas, produced from 99.8% oxygen gas with an ozone generator (Model GL-1, PCI-WEDECO, Caldwell, NJ), through a fine bubble diffuser into the test solution inside the reactor. The concentration of dissolved ozone was measured by the indigo trisulfonate colorimetric method using a molar absorptivity of 23,150 $M^{-1}cm^{-1}$ at 600 nm (Chiou *et al.*, 1995). The pH of the solution was maintained constant at a target value of 7.0 with a 0.01 M phosphate buffer. This constant pH was selected because both the kinetics of microsphere fluorescence decay and *C. parvum* inactivation were reported to be independent of pH in the range of pH 6 to 8 (Chiou *et al.*, 1997; Rennecker *et al.*, 2001).

### 2.2.2 Preozonation of YG Microspheres

Preliminary tests indicated that for both batches of YG microspheres, the initial drop of fluorescence intensity caused by ozonation was too fast for optimum simulation of *C. parvum* oocysts inactivation kinetics with the threshold concept as described in Chapter 4. Consequently, semi-batch experiments were performed to determine the fluorescence intensity decay over a wide range of ozone exposure. In this trial-and-error effort, microspheres with a dosage of 1.25 mg/L were exposed to ozone at 20 °C and pH 7. Samples were taken at *CT* intervals of 1 mg×min/L up to 25 mg×min/L, and quenched with 0.2 ml of 1.67 sodium thiosulfate solution. The fluorescence intensity of the microsphere samples was analyzed by flow cytometry as described subsequently. Fluorescence decay over the entire range of ozone exposure was examined, and efforts were made to develop correlations between fluorescence decay and *C. parvum* inactivation kinetics by the threshold concept (see chapter 4) at different ozone exposure. Such analysis revealed that optimum correlations could be achieved by reducing the fluorescence intensity of the microspheres by approximately 90 to 95%. Consequently, the fluorescence intensity was reduced by exposure to ozone in the semi-batch reactor. For Batch I, a 400 ml volume of microsphere stock with a solids concentration of 2.5% was pretreated with ozone at a temperature of 20 °C for 150 min. The ozone gas concentration was not measured but it resulted in a steady-state aqueous concentration of 1.75 mg/L when applied to 0.01 M phosphate buffer solution (pH 7). Similarly, a 400 ml volume of microsphere stock from Batch II, also having a solids concentration of 2.5%, was pretreated with ozone at a temperature of 20 °C for 200 min. The ozone gas concentration resulted in a steady-state aqueous concentration of 1.08 mg/L when applied to 0.01 M phosphate buffer solution (pH 7).

### 2.2.3 Semi-batch Ozonation Experiments

Bench-scale semi-batch ozonation experiments were performed with the Rhodamine WT and preozonated YG microspheres. Preliminary experiments were performed to assess the roles of pH (6-8), and temperature (10-25°C) in the reaction of ozone with Rhodamine WT microspheres. Additional experiments were performed with both Rhodamine WT and preozonated YG microspheres at the temperature of the raw water at the time of the full-scale demonstration as described subsequently. One mL samples were withdrawn from the reactor at *CT* intervals of 1 mg×min/L, and quenched with 0.2 ml of 1.67% sodium thiosulfate solution. The fluorescence intensity of the microspheres was analyzed by flow cytometry as described subsequently.

### 2.2.4 Batch Experiments

Batch experiments were performed to determine the kinetic of bromate formation in ACWD raw water. A schematic of the batch reactor set up is shown in Figure 2.1. Ozone gas was generated with pure oxygen fed into an ozone generator (Model GL-1, PCI-WEDECO, Caldwell, NJ) at a rate of approximately 3.5 SCFM. The gas was first bubbled through distilled de-ionized water in a gas-washing cylinder. The cleaned, humidified gas was then directed to a dilute sulfuric acid solution (pH $< 5$) in the stock cylinder. A trial-an-error effort was made to obtain the desired stock ozone concentration by adjusting the applied voltage of the ozone generator. The ozone concentration of ozone stock solution and water samples was measured by the indigo colorimetric method described by Chiou *et al.* (1995). An experiment was initiated by transferring 10 mL of the stock solution into a 100 mL gas-tight syringe containing 90 mL of ACWD raw water. The experimental temperature was maintained constant at the desired temperature by immersing the syringe in a circulating water bath. The first sample was taken with a gas-tight syringe after 1.33 min, and subsequent samples were taken at 1 min intervals. The concentration of bromate was measured by anion chromatography with post-column method as described later in the report.

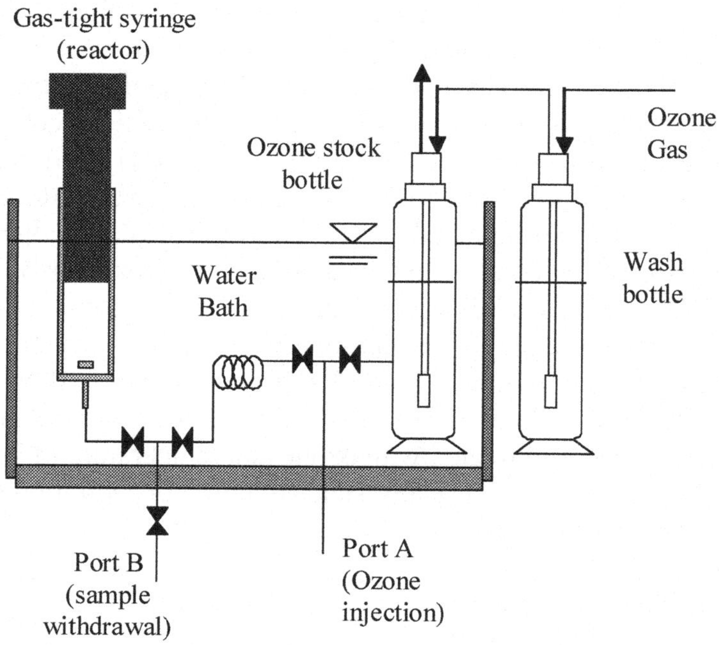


**Figure 2.1: Schematic of experimental batch apparatus**

## 2.3 FULL-SCALE DEMONSTRATION EXPERIMENTS

### 2.3.1 Ozone Contactor

Experiments were performed with one of the two parallel ozone contactors at the Alameda County Water District (ACWD) treatment plant located in Fremont, California. A longitudinal cross-section of the unit is shown in Figure 2.2. The contactor consisted of five chambers, with each chamber further divided into down-flow and up-flow sections. In addition to the dimensions shown in Figure 2.2, the contactor is 14.75 ft (4.5 m) wide, with a water column height of 22 ft (6.7 m). Water and microspheres samples were withdrawn from the sampling ports indicated in Figure 2.2. The inner ends of the sampling lines reached a distance of 7.25 ft (2.21 m) from the wall and were bent upward to avoid bubble intrusion. Ozone gas could be applied into individual chambers through porous bubble diffusers located at the bottom of the corresponding down-flow sections. This ozonation process was followed by conventional clarification treatments including coagulation, flocculation, sedimentation, and filtration.

Particle removal by these processes reduces the possibility of microspheres entering the distribution system.

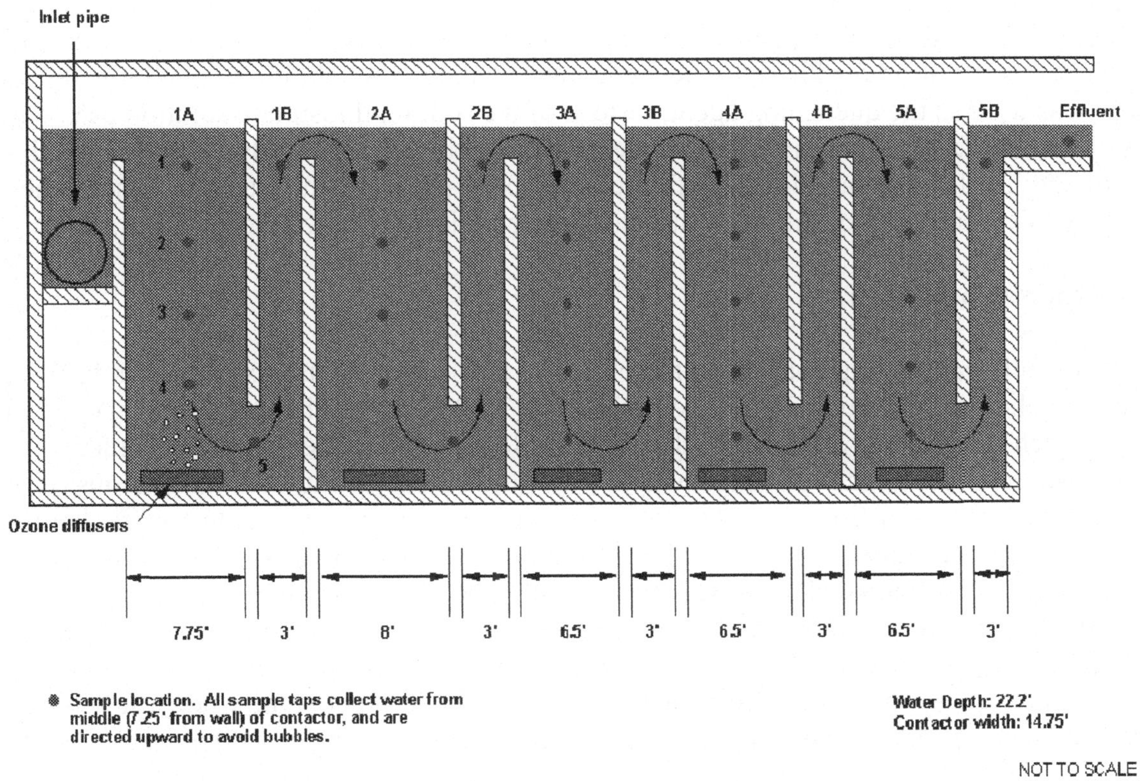


**Figure 2.2: Schematic of the full-scale ozone contactor at the Alameda County Water District Treatment Plant, Fremont, California**

### 2.3.2 Testing Protocol

The following protocol was used for Phase I of full-scale demonstration experiments. A microsphere suspension of 0.3 g/L was pumped at a flow rate of 100 ml/min into an injection port located at the influent water pipe of the ozone contactor. This resulted in a target microsphere dose of 1.0-1.2 μg/L, or 1.8-2.2 x $10^6$ microspheres/L, at the water flow rate range of 9.4-11.3 MGD at which the contactor was operated during Phase I experiments. Immediately after the addition of the microsphere suspension, the distribution of dissolved ozone concentration inside the contactor was characterized by the Indigo colorimetric method (APHA *et al.*, 1998) with samples taken in triplicate from every sampling port except Chamber 2-B, since the port was not functional. After waiting for a period of approximately twice the hydraulic retention time of the contactor, microsphere samples were taken from selected sampling ports, and then the addition of microspheres was stopped. Each sample was collected into two 1-liter containers, mixed with excess sodium thiosulfate solution to quench any residual ozone, and shipped to the University of Illinois. Water samples were also taken from the effluents of the plant sedimentation basin and filter to demonstrate the microsphere removal efficiencies. The samples were then filtered twice, first through 35 μm filters, then through 20 μm filters, to remove large particles, and the filtrate was centrifuged at 8,350×g for 30 min (Model GS-3, Sorvell) in order to retrieve the particles. The samples were analyzed by flow cytometry as described subsequently. The procedure for Phase II of the full-scale demonstration was similar to that used for Phase I with some modifications. A solution of ethyldimethyl amine (EDA) was used as the quenching agent, instead of the sodium thiosulfate solution used during Phase I, to avoid interferences by thiosulfate ion during the bromate analyses as described subsequently. Also, the samples were filtered only once using 20 μm filters because of the relatively lower turbidity of the raw water during the period of Phase II testing.

### 2.3.3 Experimental Matrix

Phase I of full-scale testing consisted of a total of twelve experiments, six each with R-WT and YG microspheres, performed during the period July 24-28, 2001. The experimental conditions used for each type of microsphere are shown in Table 2.2. Experiments were designed to investigate the effect of ozone dosage, which was varied by adjusting both the gaseous ozone concentration and water flow rate. The total ozonated gas flow rate was split equally between the first and second chamber. An overall *CT* parameter was calculated by the summation of the *CT* parameter of individual chambers, which was obtained by multiplying the average ozone concentration within each individual chamber by the fraction of the overall $T_{10}$ corresponding to that chamber ($T_{10}$ is the characteristic time for the contactor corresponding to the time when the tracer dye concentration in the effluent reached 10% of the influent value when a step input was applied). Water quality parameters representative of the raw water during that week were a turbidity of 19 NTU, and a total organic carbon (TOC) concentration of 1.9 mg/L.

As discussed in a subsequent section, flow cytometry analyses of the Rhodamine WT samples revealed that the fluorescence of the microspheres was not distinguishable from that of background particles of similar size. Accordingly, Phase II experiments were performed using only YG microspheres. Phase II of full-scale testing was performed during the period May 28-30, 2002. The experimental conditions for the six experiments performed are presented in Table 2.3. The effect of variable ozone dosage was investigated by varying both the gaseous ozone concentration, and the water flow rate. The high water flow rate was achieved by bypassing the second contactor and passing the entire flow through the experimental unit. Ozonated gas was applied only to the first chamber for Tests Nos. 1 to 5, and split between chambers 1 and 2 for Test No. 6. Ozonated gas was dosed in the first and second chambers for Test No. 6 because the water flow rate was too high to achieve the target *CT* value if the ozonated gas was applied only to the first chamber. Water quality parameters representative of the conditions encountered during Phase II experiments were a turbidity of 8 NTU, a total organic carbon (TOC) concentration of 3.5 mg/L, and a bromide concentration of 150 μg/L.

**Table 2.2: Experimental matrix for Phase I of full-scale testing**

| Test | Date | Water Flow Rate (MGD) | Gas Flow Rate (scfm)* | Temperature (°C) | Feed Gas Conc. (% wt.) | Ozone Dosage (mg/L) | *CT*** (mg×min/L) |
|---|---|---|---|---|---|---|---|
| 1 | 7/24/01 | 11.0 | 67/66 | 22.3 | 2.38 | 3.58 | 1.56 |
| 2 | 7/25/01 | 11.1 | 110/113 | 22.9 | 1.87 | 4.79 | 4.20 |
| 3 | 7/25/01 | 11.3 | 70/71 | 22.9 | 1.92 | 3.00 | 0.87 |
| 4 | 7/26/01 | 9.4 | 108/107 | 23.4 | 1.63 | 4.79 | 5.27 |
| 5 | 7/26/01 | 9.4 | 64/62 | 23.4 | 1.68 | 2.68 | 0.42 |
| 6 | 7/26/01 | 9.4 | 86/84 | 23.2 | 1.66 | 3.66 | 2.15 |

* First/second chambers

** Product of experimentally determined average ozone concentration and $t_{10}$

**Table 2.3: Experimental matrix for Phase II of full-scale testing**

| Test | Date | Water Flow Rate (MGD) | Gas Flow Rate (scfm)* | Temperature (°C) | Feed Gas Conc. (% wt.) | Ozone Dosage (mg/L) | *CT*** (mg×min/L) |
|---|---|---|---|---|---|---|---|
| 1 | 5/28/02 | 9.5 | 106/0 | 18.9 | 1.87 | 2.65 | 0.91 |
| 2 | 5/29/02 | 9.7 | 126/0 | 18.9 | 2.58 | 4.20 | 3.73 |
| 3 | 5/29/02 | 9.5 | 124/0 | 19.1 | 2.02 | 3.29 | 1.75 |
| 4 | 5/30/02 | 15.5 | 120/0 | 19.7 | 3.19 | 3.18 | 1.53 |
| 5 | 5/30/02 | 15.5 | 120/0 | 19.7 | 2.87 | 2.83 | 1.27 |
| 6 | 5/30/02 | 15.5 | 120/119 | 20.3 | 2.03 | 4.03 | 2.74 |

* First/second chambers

** Product of experimentally determined average ozone concentration and $t_{10}$

### 2.3.4 Full-scale Testing Precautions

In one of the tests in phase I testing, two water samples were withdrawn from the influent of the ozone contactor before the testing. During the testing, in addition to withdrawing the sample from the contactor effluent for the microspheres analysis, two samples were also taken from the effluents of the sedimentation tank and filter. The samples were centrifuged to concentrate the solids concentration by a factor of about 500 times, and then analyzed by flow cytometry. Figure 2.3 shows the results of the flow cytometry analyses for the water at the various locations sampled. The x and z axis represents the relative fluorescence intensity and the number of particles with the corresponding intensity, respectively. As depicted from the figure, the (blank) raw water contained some particles, which were also present in the contactor effluent. Two peaks were observed for the contactor effluent, with the peak with greater fluorescence denoting the microspheres population, and that with lower fluorescence representing natural particles. The latter particles were also observed in the raw water sample. Nevertheless, the number of all particles decreased drastically as the water passed through the sedimentation tank and the filter, as shown by the decrease or disappearance of the peaks in the respective effluent samples. Table 2.4 shows the number of particles in the various samples that have the fluorescence intensity between 10 and 40 (the fluorescence intensities of the microspheres population as depicted from Figure 2.3) during the experiment. Based on Table 2.4, it can be seen that the removal efficiency of the sedimentation tank plus the filter is at least 99%. With the target concentration of microspheres of 1 μg/L during testing, it is anticipated that the mass concentration of microspheres at the effluent will be less than 0.01 μg/L, which means that the dye concentration will be less than 2.7 ng/L. It is important to point out that this calculation gives the worst-case scenario, which is based on the estimation that the weight percent of dye in microspheres is 27%. The actual percentage, and thus the actual dye concentration in water, should be somewhat lower. Therefore, it is believed that the testing should not exert any negative public health impact.

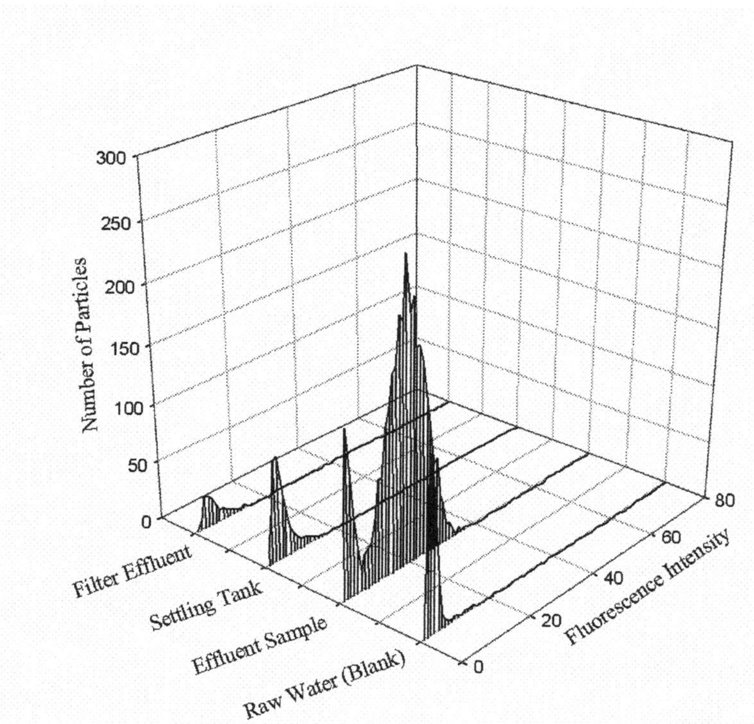


**Figure 2.3: Flow cytometry analysis of samples taken during Phase I of full-scale experiments**

**Table 2.4. Number of particles in water withdrawn at different locations with fluorescence intensity within the range of 10 and 40 during the previous full-scale experiment**

| Sampling Location | Number of Microspheres | Fraction Remaining (No./No. in Contactor Effl.) |
|---|---|---|
| Contactor Effluent | 2952 | --- |
| Settling Tank Effluent | 56 | 0.019 |
| Filter Effluent | 24 | 0.0081 |

## 2.4 ANALYTICAL METHODS

### 2.4.1 Flow Cytometry

The distribution of fluorescence intensity of Rhodamine WT microspheres was analyzed with a MoFlo MLS flow cytometer (Cytomation, Fort Collins, CO). The samples were exposed to a laser beam set at 250 mW. Approximately 14,000 events, which included both particles and background signal, were analyzed per second, and three measurements were made for each particle: forward light scatter (FSC) at 448 nm, side light scatter (SSC) at 448 nm, and fluorescence emission at 558 nm. Interference by most extraneous particles was eliminated by "gating" the forward scattering (FS) and side scattering (SS) signals within narrow ranges corresponding to those for the microspheres. The results were analyzed using the flow cytometer system software.

The distribution of fluorescence intensity of YG microspheres in experimental samples was measured with a Coulter Epics XL-MCL flow cytometer (Beckman-Coulter, Fullerton, CA). The machine was equipped with an air-cooled argon ion laser operating at a wavelength of 488 nm with the output adjusted to 15 mW. The fluorescence emission of individual microspheres was measured at the wavelength of 525 nm. The relative fluorescence scale, ranging from 1 to 1,024, was adjusted by adjusting the detector voltage and gain values so that the average fluorescence of the pretreated stock was about 54 for Phase I, and 70 for Phase II. The corresponding average fluorescence of the untreated microsphere stock was about 850 for both phases.

### 2.4.2 Bromide and Bromate ions

Prior to analysis, bromide and bromate samples were filtered with 0.45-μm syringe filters. Bromide ion was analyzed with an ion chromatograph (Model DX-300, Dionex, Sunnyvale, CA) using conductivity detection and the following components and conditions: IonPac® AG9-HC (4×50 mm, Dionex) guard column; IonPac® AS9-HC (4 x 250 mm, Dionex) analytical column; 9 mM $Na_2CO_3$ eluent; 1.0 mL/min eluent flow rate; 500 μL sample loop size; PED-2 (Dionex) detector; ASRS-Ultra (Dionex) suppressor; external water suppression mode.

Bromate ion was also measured by ion chromatography, but instead of using the conductivity detector, a post-column derivatization method (Echigo *et al.*, 1999), a modified version of the method developed by Weinberg and Yamada (1998), was used. The following components and conditions were used: guard column IonPac® AG9-HC (4×50 mm, Dionex); IonPac® AS9-HC (4 x 250 mm, Dionex) analytical column; 9 mM $Na_2CO_3$ eluent; 1.0 mL/min eluent flow rate; 500 μL sample loop size; 500 mg/L ODA, 5g/L NaBr, 200 g/L methanol, and 5.6% $HNO_3$ post-column reagent; 0.4 mL/min reagent flow rate; 55 °C reaction temperature; UV/VIS detector (LINEAR Model 200, Alltech, IL).

## 2.5 MATHEMATICAL MODEL

### 2.5.1 *C. parvum* Oocyst Inactivation Kinetics

At the initial phase of this study, the kinetics of *C. parvum* oocyst inactivation with ozone obtained by Rennecker *et al.* (1999) was chosen. Studies performed by Rennecker *et al.* (2001), Corona-Vasquez *et al.* (2002) and Hirata *et al.* (2001) showed consistency between data obtained from modified in-vitro excystation method and animal infectivity method. The kinetics is characterized by an initial lag phase during which little or no inactivation occurs followed by a pseudo first-order rate law. The delayed Chick-Watson model (Equation 2.1) proposed by Rennecker *et al.* (1999) was used for Phase I of the study:

$$\frac{N}{N_0} = \begin{cases} 1 & \text{if } CT \le CT_{lag} = \frac{1}{k_N}\ln\left(\frac{N_1}{N_0}\right) \\ \frac{N_1}{N_0}\exp(-k_N Ct) = \exp(-k_N\{CT - CT_{lag}\}) & \text{if } CT > CT_{lag} = \frac{1}{k_N}\ln\left(\frac{N_1}{N_0}\right) \end{cases} \quad (2.1)$$

where $N/N_0$ = survival ratio; $N_1/N_0$ = lag phase factor; $k_N$ = pseudo first-order inactivation rate constant [L/(min×mg)]; $CT_{lag}$ = $CT$ required to overcome the initial lag phase [mg×min/L]. The parameters that determine the inactivation kinetics, such as $N_1/N_0$ and $k_N$, can be obtained from fitting experimental data obtained with a semi-batch reactor in which the dissolved ozone concentration is maintained constant by continuously bubbling of ozonated gas (Rennecker *et al.*, 1999).

During the course of this study, Clark *et al.* (2002) examined four sets of inactivation kinetics (Rennecker *et al.*, 1999; Li *et al.*, 2000; Owens *et al.*, 2000; Oppenheimer *et al.*, 2000) and proposed the following *Ct* equation:

$$Ct = 8.5 \times (1.10)^{-T} \times (-Ln(N_t / N_0)) \quad (2.2)$$

Where $Ct$ = ozone exposure required to achieve a certain level of inactivation [mg×min/L]; $T$ = temperature [°C]; $N_t/N_0$ = survival ratio. This equation is currently under consideration for the development of $C \times T$ requirements under the LT2ESWTR. Accordingly, the microspheres for

Phase II of this study were optimized, as described in Section 2.2.2, to match the inactivation kinetics corresponding to Equation 2.2.

### 2.5.2 Modeling of Ozone Decomposition

The decomposition of ozone in natural water can be described by the following differential equations (Kim *et al.*, 2002a,b):

$$\frac{dC}{dt} = -k_D CD - k_s C \tag{2.3}$$

$$\frac{dD}{dt} = -k_D CD \tag{2.4}$$

where $C$ = ozone concentration [mg/L]; $D$ = fast demand concentration [mg/L as $O_3$]; $k_D$ = second-order rate constant for the reaction between ozone and fast demand [L/mg-min]; and $k_s$ = pseudo first-order rate constant for the decomposition of ozone in natural water [$min^{-1}$]. The initial demand was assumed to be relatively fast, so the second-order reaction kinetics $k_D$, was set to be a relatively high default value of $1.6 \times 10^4$ M/mg-min. The other parameters, i.e., $k_s$ and initial fast ozone demand $D_0$, were obtained by fitting batch-ozonation data. The parameters were then incorporated into a mathematical model designed to predict the performance of full-scale ozone contactors as described in the following section.

### 2.5.3 Modeling of Full-Scale Ozone Contactor

Each chamber of a multi-chamber bubble diffuser contactor can be modeled as an axial dispersion reactor (ADR), in which a single parameter (i.e., dispersion number) is used to represent the non-ideal hydrodynamics. In general, the steady-state mass balance for each chemical species (e.g., dissolved ozone, bromate), or microorganism can be represented by the following second-order non-linear ordinary differential expression (Kim *et al.*, 2002a,b):

$$0 = d\frac{d^2C_i}{dz^2} \pm \frac{dC_i}{dz} + \theta \cdot f_i(C_1, C_2, \cdots, C_N) \qquad i = 1, 2, \cdots, N \tag{2.5}$$

where $d$ [dimensionless] = dispersion number; $z$ [dimensionless] = normalized downward distance from the water surface; and $\theta$ [T] = hydraulic residence time in the chamber. In the case of a bubble-diffuser column, interphase mass transfer terms are included as part of $f_i$ for gas and liquid phase ozone as described in Mariñas *et al.* (1993). The sign in the second term of Equation 2.5 is determined by the chamber operation mode (i.e., negative for counter-current or positive for co-current in the case of the liquid phase). In this model, it was assumed that $d = 0$ in the gas phase (thus simplifying to a first-order differential equation) and gas volume fraction throughtout the bubble-diffuser column was negligible. The above set of second order non-linear ordinary differential equations is subject to *2N-1* boundary conditions of the form (Kim *et al.*, 2002a,b):

$$C_i\big|_{inlet} = C_{i,in} \pm d\frac{dC_i}{dz}\bigg|_{inlet} \qquad i = 1,2,\cdots,N$$

$$\frac{dC_i}{dz}\bigg|_{outlet} = 0 \qquad i = 1,2,\cdots,N \tag{2.6}$$

Consistent with the signs in the second term of Equation 2.5, the sign of the dispersion term in the first boundary condition part of Equation 2.6 is positive for counter-current and negative for co-current operation mode.

Each second order ordinary differential equation except that for the gas phase was replaced by a set of two first-order ordinary differential equations by introducing the auxiliary variable $C_i^*$ defined as (Kim *et al.*, 2002a,b)

$$\frac{dC_i}{dz} = C_i^* \tag{2.7}$$

Equation 2.5 can then be converted into the following equivalent expression (Kim *et al.*, 2002a,b):

$$\frac{dC_i^*}{dz} = \pm\frac{1}{d}\cdot C_i^* - \frac{1}{d}\cdot\theta\cdot f_i(C_1, C_2, \cdots, C_N) \tag{2.8}$$

The resulting set of *2N-1* first-order differential equations was substituted with finite-difference expressions on a mesh of points that covered the range of integration. An iterative calculation was performed until the deviations from the finite-difference approximation of each equation at each mesh point were sufficiently small (Kim *et al.*, 2002a,b). The calculation algorithm, called the relaxation method, was adopted mostly from Press *et al.* (1992) and modified and programmed with C++ programming language (Kim *et al.*, 2002a,b).

# CHAPTER 3. DEVELOPMENT OF RHODAMINE WT MICROSPHERES

An initial goal of this project was an attempt to develop a microsphere manufactured with a dye certified for use in drinking water. The fluorescent dye Rhodamine WT was specifically chosen because it has been certified by NSF (ANSI/NSF, 2000) and reviewed by the EPA for use in drinking water. Rhodamine WT, chemically different from other Rhodamine dyes, was developed specifically for water tracing work (U.S. patent 3,367,946). It was initially included on the 1999 draft Contaminant Candidate List (CCL), but it was subsequently removed from the final CCL version because the EPA concluded that it would have no adverse public health effects if NSF recommendations were followed (USEPA, 1998b).

One major challenge with using Rhodamine WT as the dye is its relatively high solubility in water. Unfortunately, this property that makes it an excellent chemical for performing tracer tests in natural and engineered systems, also results in Rhodamine WT leaching out from the polystyrene matrix of commercial microspheres. Faced with this dilemma, consultation with research and development staff from Bangslabs, Inc. resulted in a potentially viable approach. The leaching of Rhodamine WT could be avoided by binding the carboxylic group of the dye molecule to amine groups on the surface of modified polystyrene microspheres. Unfortunately, one potential weakness was that the overall mass of dye and corresponding fluorescence of the resulting surface-labeled microspheres would be much lower than those of commercial particles manufactured with hydrophobic dyes, such as the YG dye used in this study, which penetrated into at least a portion of the polystyrene matrix. Nevertheless, the task of developing the Rhodamine WT microspheres was undertaken.

The developmental approach consisted of two stages. During the first stage, six lots of microspheres prepared by immersing the microspheres into solutions with a wide range of dye mass concentrations were prepared and tested with the goal of providing a gross estimate of the appropriate mass of dye needed. In the second stage, microspheres were manufactured with solutions containing a mass of dye within a narrower range for the purpose of final optimization. The resulting optimal Rhodamine WT microspheres were tested during Phase I of the full-scale experiments, but unfortunately, fluorescent background particles present in the raw water and having similar size to that of the microspheres interfered with the flow cytometry analyses of the samples. Of course, the next step would have been to develop a Rhodamine WT microsphere with the dye cross-linked not only to the external surface but also internally. Unfortunately, the limited resources available for this unsolicited project did not allow further development and optimization of this microsphere. Because of the potential for such development in the future, the bench-scale experiments performed with the Rhodamine WT microspheres are reported in this section. These data could be used as the starting point for future development of improved surrogates. Because of the background interferences, Rhodamine WT microspheres were not used during Phase II of the full-scale testing.

## 3.1 PRELIMINARY TESTING OF RHODAMINE WT MICROSPHERES

### 3.1.1 Flow Cytometry Analysis of Rhodamine WT Microspheres

Representative flow cytometry results obtained with control with Rhodamine WT microsphere samples (i.e., stocks of microspheres unexposed to ozone) are shown in Figures 3.1 and 3.2 for the case of microsphere stock #3. The plot shown in Figure 3.1 is a distribution of forward light scatter (FSC) and side light scatter (SSC) for all particles present in the sample. FSC and SSC are relative measurements of particle size and material light scatter properties, respectively. For example, the FSC value of 10 corresponds to a particle size of approximately 1 micrometer. Two distinct regions can be observed in Figure 3.1. The region with SSC below approximately 7 corresponded to background signal also observed with blank samples. The second region at higher SSC corresponded to Rhodamine WT microspheres from stock #3.

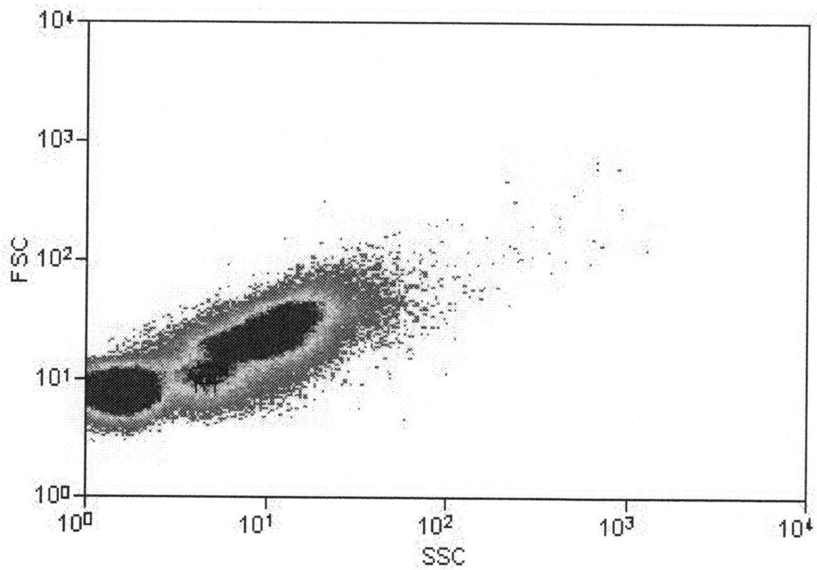


**Figure 3.1: Forward Light Scatter (FSC) vs. Side Light Scatter (SSC) distribution obtained from flow cytometry analysis of microsphere stock #3 (unexposed to ozone)**

The plot in Figure 3.2 shows the fluorescence intensity distribution for the same sample of stock #3 shown in Figure 3.1. The two peaks at low fluorescence intensity corresponded to background signal also observed with blank samples, essentially corresponding to the background signal region in Figure 3.1. The remaining peaks at high fluorescence intensity corresponded to Rhodamine WT microspheres from stock #3. The multiple peaks revealed the occurrence of particle clumping. It is interesting to observe that the maximum fluorescence intensity of the peaks in the intervals or gates R2 and R3 shown in Figure 3.2 are approximately twice and three times, respectively, of that for the peak in gate R1 consistent with the presence of single particles (R1) and clumps of two (R2) and three (R3) particles. Figure 3.1 also revealed the presence of clumps with two regions with size approximately twice and three times compared to those in gate R1 (note: gate R1 in Figure 3.1 is not exactly the same as gate R1 in Figure 3.2

as discussed subsequently) being observed, though not as clearly as with the fluorescence intensity plot.

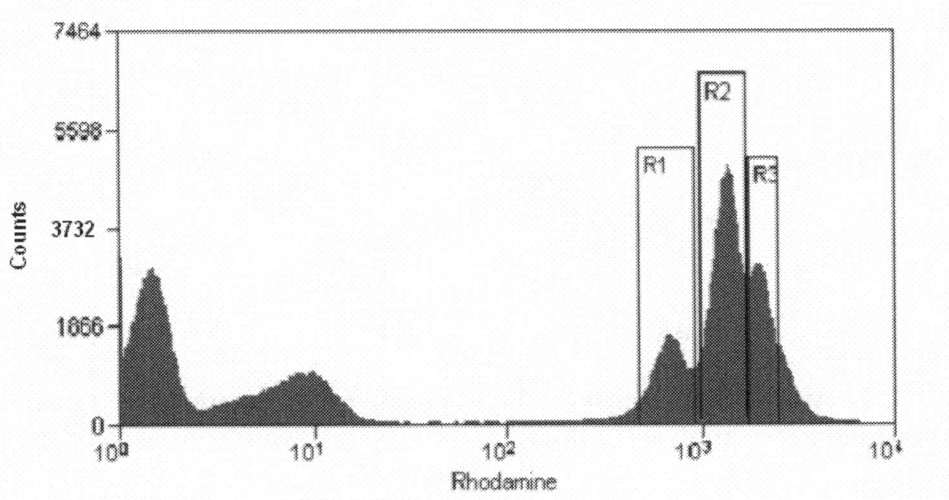


**Figure 3.2: Fluorescence distribution obtained from flow cytometry analysis of microsphere stock #3 (unexposed to ozone)**

A potential problem associated with the flow cytometry results is that the signals from single particles, clumps and background signal (Figure 3.1), and single particles and clumps (Figure 3.2) can overlap. The flow cytometry data analysis software was used to overcome this problem, and also to demonstrate the correspondence between particles shown in Figures 3.1 and 3.2. For example, the gate R1 in Figure 3.2 was selected to obtain the corresponding FSC versus SSC plot shown in Figure 3.3. Essentially all particles with fluorescence intensity in gate R1 of Figure 3.2 are the single particles of Figure 3.1. However, because of variability in both FSC and SSC associated with the flow cytometry analysis, some of the single particles would overlap with the background signal as well as with some of the clumps as it could be demonstrated by superimposing Figures 3.1 and 3.3.

A more useful approach was to first gate the center of the single particle region in Figure 3.1 (gate R1), and then to look at the fluorescence intensity for just these particles. The corresponding fluorescence intensity distribution is shown with a single peak in Figure 3.4. Although this approach will result in excluding some of the single microspheres, it will enable the study of the dynamics in both control samples and samples exposed to ozone at various *CT* levels without interference from clumps or background signal.

Similar experiments were performed with all other microsphere stocks and blank. Only stocks #1, #2 and #3 were found to emit fluorescence at levels that could be distinguished from those by the blank, and therefore, Stocks #4, #5 and #6 were dropped for further study. The average fluorescence intensity emitted by microspheres from stocks #1 and #2 were lower compared to that for stock #3 microspheres due to an auto-quenching effect at the relatively high concentrations of Rhodamine WT in microspheres from stocks #1 and #2. This effect will be discussed in more detail in the following section.

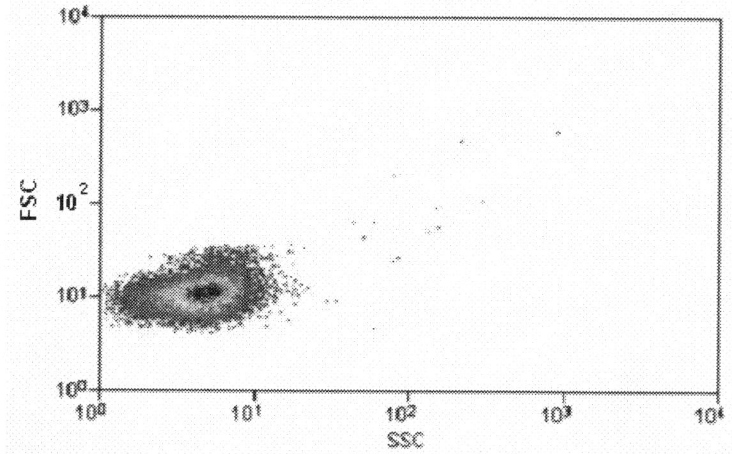


**Figure 3.3: Forward Light Scatter (FSC) vs. Side Light Scatter (SSC) distribution obtained from flow cytometry analysis of microsphere stock #3 (unexposed to ozone) within gate R1 of Figure 3.2**

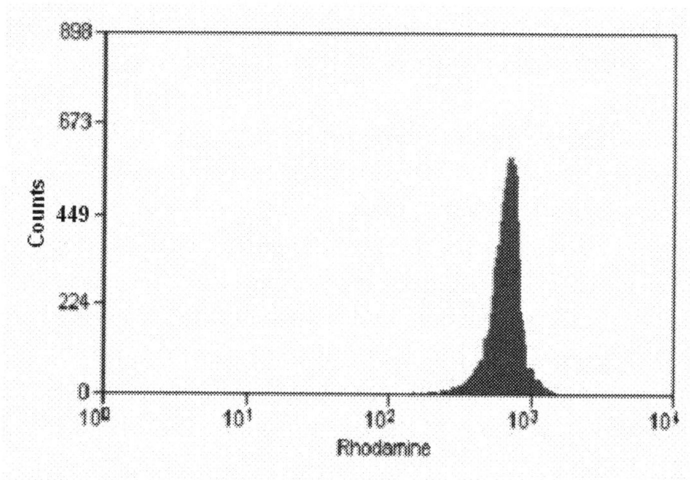


**Figure 3.4: Fluorescence distribution obtained from flow cytometry analysis of microsphere stock #3 (unexposed to ozone) within gate R1 of Figure 3.1**

### 3.1.2 Semi-batch Ozonation Experiments

The effect of exposure to ozone on fluorescence intensity distribution of single microspheres (gate R1) from stocks #1, #2 and #3 are shown in Figures 3.5 (a), (b) and (c), respectively. The control samples for all three stocks had a single peak with relatively high fluorescence efficiency. Increasing exposure to ozone resulted in a general decrease in the average intensity of the peak and the appearance of a second peak of lower intensity, which became predominant at the highest *CT* values investigated. The second peak corresponded to the fluorescence intensity emitted by the polystyrene matrix of the particles. The average fluorescence intensities as a function of *CT* are summarized in Table 3.1.

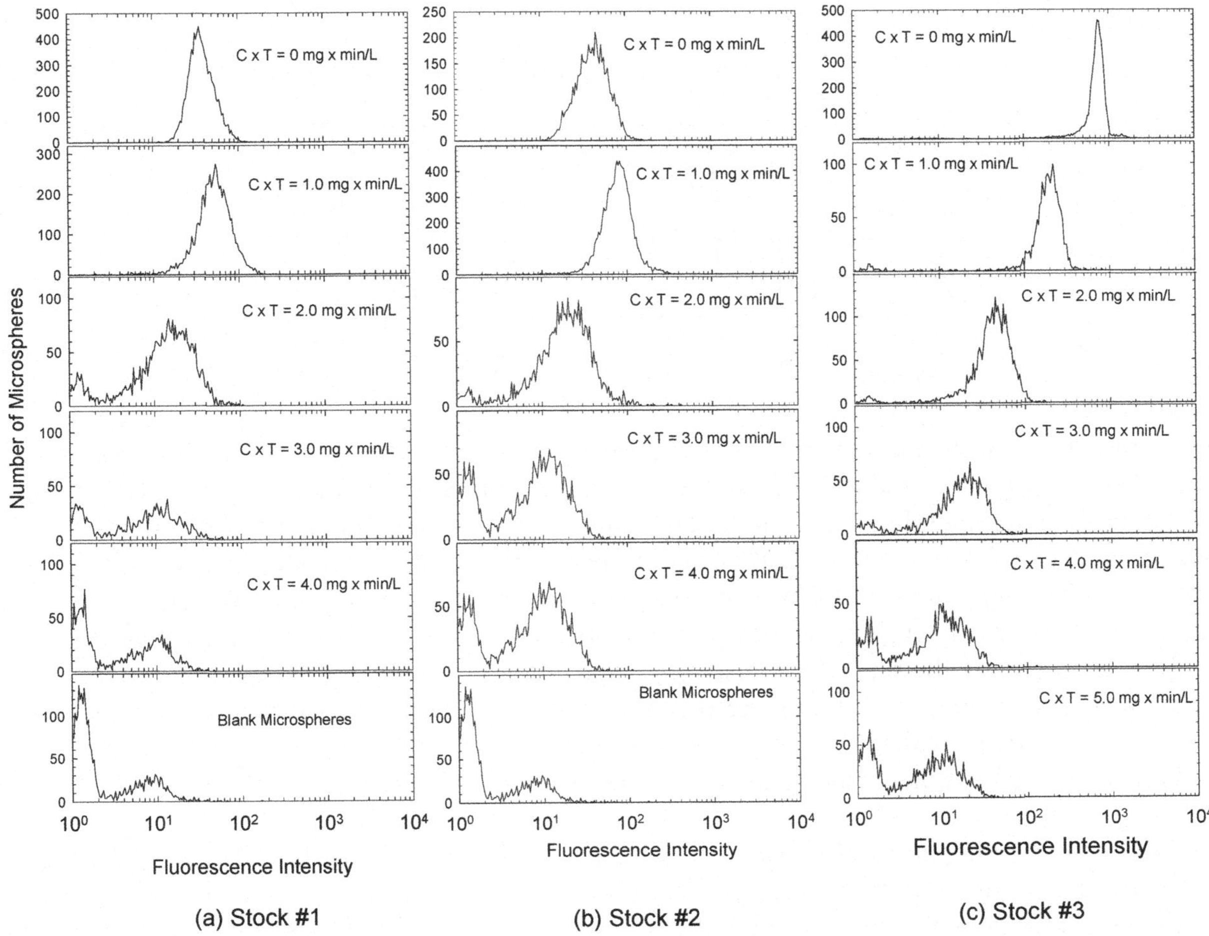


**Figure 3.5: Fluorescence distribution obtained from flow cytometry analysis of microsphere (a) stock #1 (b) stock #2 and (c) stock #3 within gate R1 exposed to ozone at the indicated *CT* levels (pH 7, 20 °C)**

An interesting observation is that the control samples for microspheres from stocks #1 and #2 (Figure 3.5 (a) and (b)) have much lower fluorescence intensity than microspheres from stock #3 even though the Rhodamine WT mass used in the preparation of stocks #1 and #2 was $10^4$ and $10^2$ times, respectively, that for stock #3. These results are consistent with the well known phenomenon of fluorescence quenching, i.e., excited electron returning to the ground state via a non-radiative pathway, which occurs at relatively high concentrations of fluorescence molecules. Surrounding molecules, if present at high concentration, can act as electron acceptors for non-radiative transfer (Shapiro, 1995). This was consistent with the observation in Figures 3.5 (a) and (b), and Table 3.1 that the average fluorescence intensity of the samples after an ozone exposure corresponding to a $CT = 1.0$ mg×min/L was higher than that for the control. It appeared than once some of the fluorescent chemical molecules are attacked by ozone, the quenching effect was diminished.

**Table 3.1: Summary of average fluorescence intensities for stocks #1 to #3 as shown in Figure 3.5 (a), (b) and (c)**

| $C \times T$ (mg×min/L) | Stock #1 | Stock #2 | Stock #3 |
|---|---|---|---|
| 0 | 40.7 | 45.5 | 727. |
| 1 | 56.1 | 83.7 | 193. |
| 2 | 16.2 | 22.8 | 45.0 |
| 3 | 9.35 | 9.62 | 17.8 |
| 4 | 5.77 | 6.09 | 10.2 |
| 5 | - | - | 7.23 |
| Blank | 3.59 | 3.59 | 2.9 |

The average fluorescence intensities of the peak with relatively higher fluorescence in Figures 3.5 (a), (b) and (c) were used to compute the normalized ratios presented in Figure 3.6 as a function of the corresponding *CT* values. The value *I,* or $I_o$, was obtained by subtracting the average intensity of the higher intensity peak in the corresponding blank sample (given in Table 3.1) from the average intensity of the sample or control (given in Table 3.1). As depicted in Figure 3.6, the normalized fluorescence for microspheres from stocks #1 and #2 increased initially. Interestingly the rate of fluorescence decay after a *CT* of 1 mg×min/L was approximately the same for all three microspheres stocks. Although an explanation for such results could not be provided, they might indicate that lot variability might not be as severe for Rhodamine WT microspheres as it was for YG particles (see Chapter 4).

Based on the experimental results, it appears that microspheres from stock #3 might be the most promising particles to be used as surrogate indicators for *C. parvum* oocysts due to the highest fluorescence intensity. Stocks #4 to #6 with mass of dye of $10^{-2}$–$10^{-6}$ times that used for stock #3 have average intensities at or near levels observed for the blank samples, and therefore they were dropped from further consideration.

Experimental results obtained for tests performed to assess the effect of pH (pH 6-8), and temperature (10-25°C) on the kinetics of fluorescence intensity decay of microspheres from stock #3 exposed to various levels of ozone corresponding to *CT*= 0-5 mg×min/L are presented in Figures 3.7, and 3.8, respectively. Consistent with the observation for *C. parvum* oocyst inactivation with ozone, the decay rate of fluorescence was essentially independent of pH within the range investigated. In contrast, the rate of fluorescence decay increased with increasing temperature, but this dependence was not as strong as that for *C. parvum* inactivation with ozone as discussed in the next section.

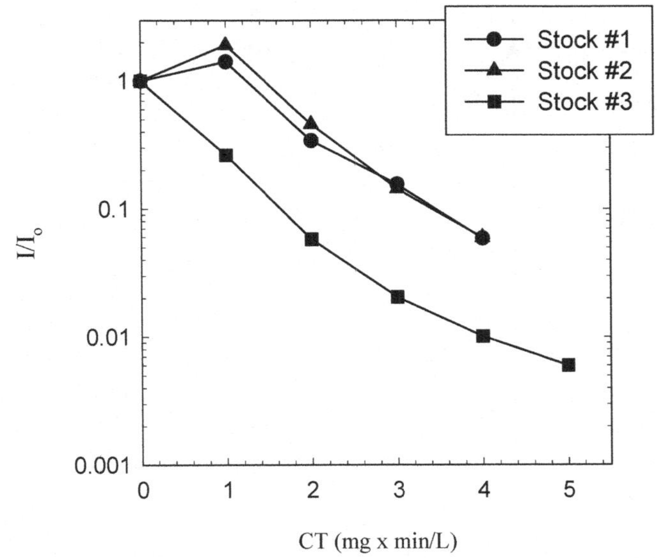


**Figure 3.6: Kinetics of average fluorescence intensity decay of microspheres from Stocks #1-3 as a Function of Ozone Exposure (pH 7, 20°C)**

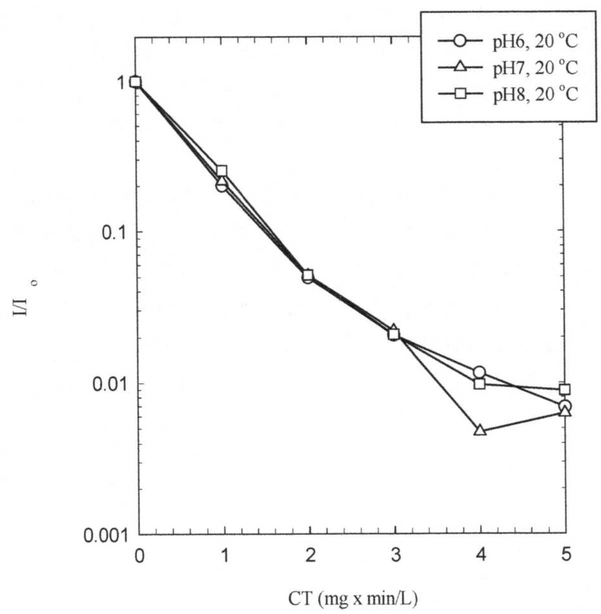


**Figure 3.7: Effect of pH on the kinetics of average fluorescence intensity decay of microspheres from Stock #3 exposed to ozone (20°C)**

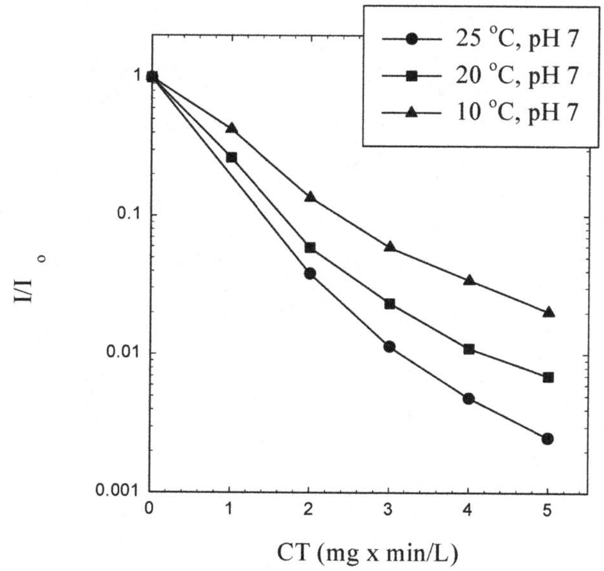


**Figure 3.8: Effect of temperature on the kinetics of average fluorescence intensity decay for microspheres from stock #3 exposed to ozone**

### 3.1.3 Comparison of Fluorescence Decay and C. parvum Inactivation Kinetics

Efforts made to use the threshold concept (Chiou *et al.* 1997), used for YG microspheres in Chapter 4, proved unsuccessful primarily because of a relatively faster decrease in fluorescence intensity observed with the Rhodamine WT microspheres. In contrast, the average fluorescence intensity decay approach, described in the preceding section, appeared to provide kinetics more comparable to that reported for *C. parvum* inactivation with ozone.

Comparisons of Rhodamine WT microsphere (stock #3) average fluorescence intensity decay kinetics obtained in this study and *C. parvum* oocyst inactivation data from the literature (Rennecker *et al.*, 1999, Clark *et al.*, 2002) are shown in Figure 3.9 for 20 °C, and Figure 3.10 for 10 and 25 °C. These initial comparisons reveal a reasonably good agreement between the kinetics for the ozone reactions with the non-biological surrogates and *C. parvum* oocysts. Within the range investigated, better correspondence was observed at relatively higher temperatures.

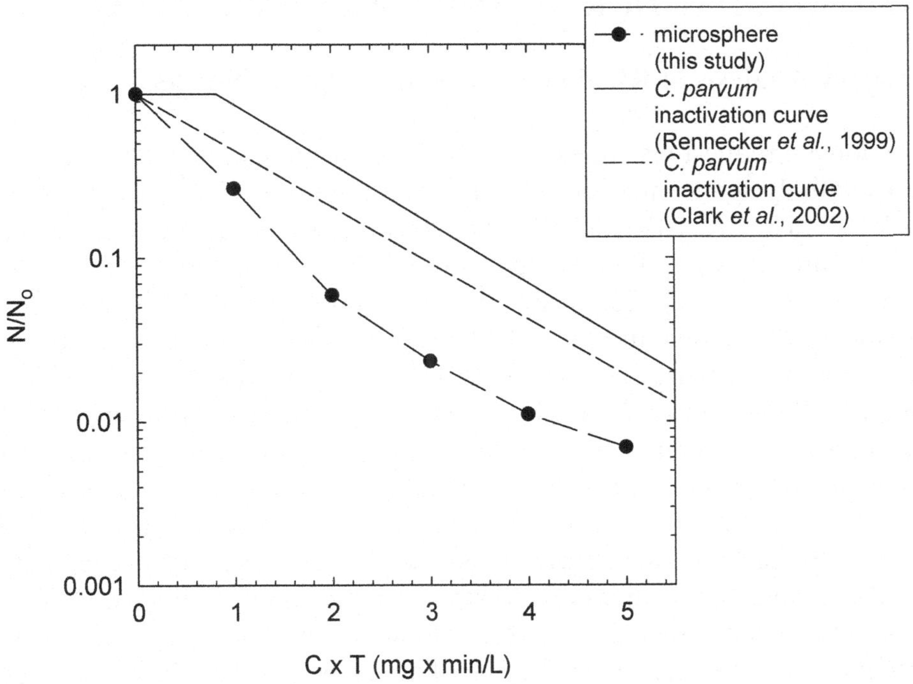


**Figure 3.9: Comparison of Rhodamine WT microsphere (stock #3) average fluorescence intensity decay and *C. parvum* oocyst inactivation kinetics with ozone at 20 °C**

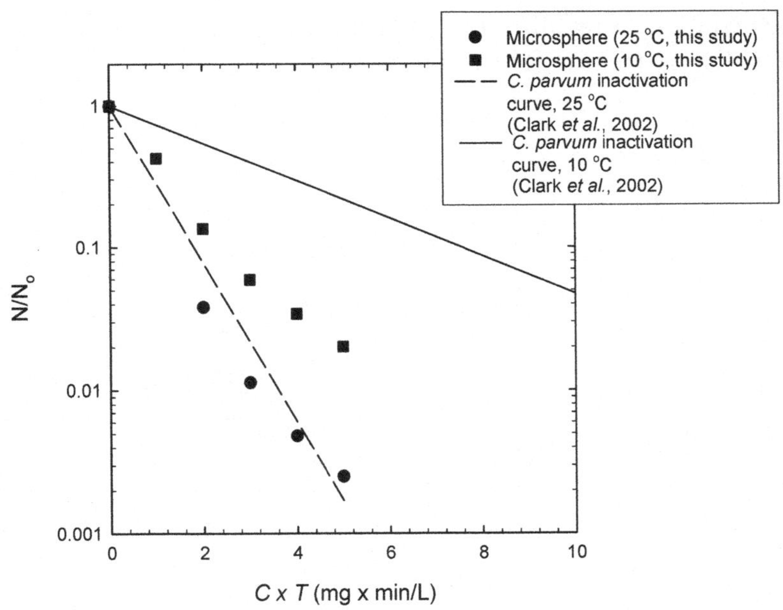


**Figure 3.10: Comparison of Rhodamine WT microsphere (stock #3) average fluorescence intensity decay and *C. parvum* oocyst inactivation kinetics with ozone at 10 and 25 °C**

## 3.2 OPTIMIZATION TESTING OF RHODAMINE WT MICROSPHERES

### 3.2.1 Flow Cytometry Analysis of Rhodamine WT Microsphere Stocks

As mentioned previously, request was made to Bangslabs, Inc. to manufacture additional microsphere stocks with dye concentrations closer to that use for stock #3. They were referred to as stocks 1/50X, 1/10X, 1X, 2X, 10X, and 50X, which corresponded to the relative amount of dyes used to manufacture stock #3. For example, stock 1/50X was manufactured with 1/50$^{th}$ times the dye compared to stock #3. Representative results obtained from flow cytometry analyses of microspheres are shown in Figure 3.11 for stock #3 from the preliminary testing and stock 1X. The plot shown in Figure 3.11 is a distribution of forward light scatter (FSC) and side light scatter (SSC) for all the signals detected by the flow cytometer. The figure reveals that the two batches have somewhat different FSC vs. SSC properties. More clumps of microspheres can be observed in stock #3, as shown in the denser population within the rectangular region (R2) for stock #3. Figure 3.12 shows the plot of FSC vs. fluorescence intensity for the two stocks. As depicted in the figure, the plot can be interpreted as two distinct populations. The populations on the left with low fluorescence intensities correspond to the background signals, which is always present even if no particles are detected. (The background signals can also be seen in Figure 3.11 as the dense population on the left, with FSC value of about $10^{0.5}$ to $10^{1.1}$ and SSC value of about $10^{0}$ to $10^{0.4}$). The population on the right corresponds to the microspheres. Figure 3.12 once again shows that stock #3 contained more clumps than stock 1X. No explanation can be provided for this difference between the previous and the new stocks. As done during the preliminary testing, only the single particles were considered for fluorescence measurement. This was accomplished by "gating" the single particle population with the aid of the flow cytometry analysis software (region R1 in Figure 3.11).

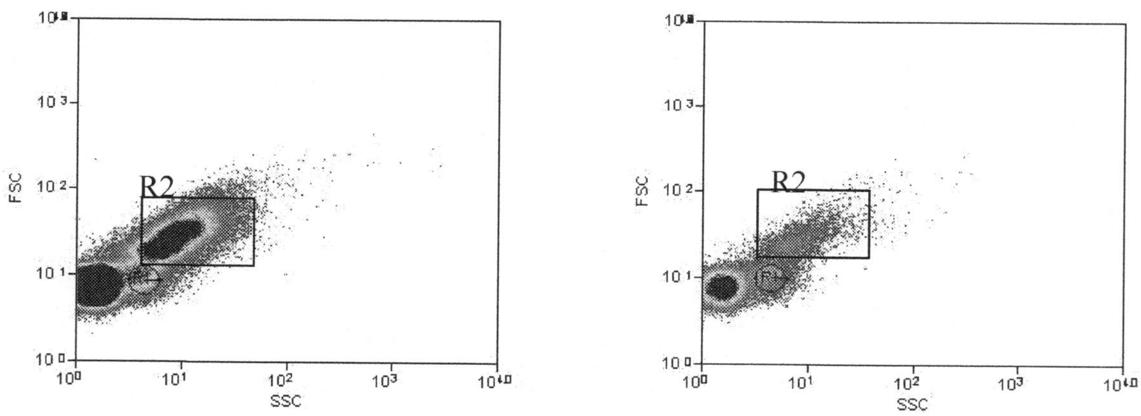


**Figure 3.11: FSC vs. SSC diagrams for stocks #3 (left) and 1X (right) obtained from flow cytometry analyses**

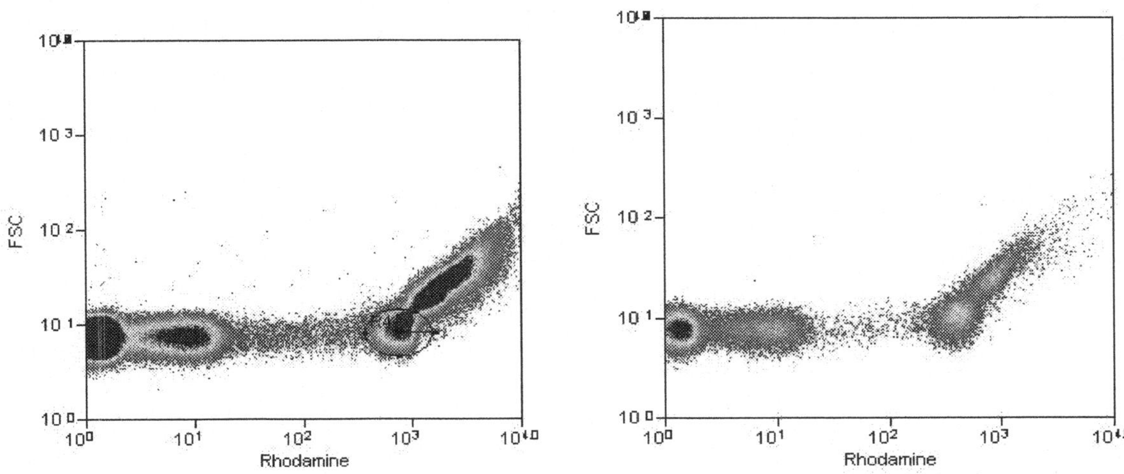


**Figure 3.12: FSC vs. fluorescence intensity plots for stocks #3 (left) and 1X (right) obtained from flow cytometry analyses**

The fluorescence intensity distributions within R1 in Figure 3.11 for the different stocks are shown in Figure 3.13, and the corresponding average intensities are listed in Table 3.2. Among the new stocks, stock 1X gave the highest fluorescence intensity, despite the fact that less dye was used compared to stocks 10X and 50X. This phenomenon was consistent with the quenching effect (Shapiro, 1995) discussed in the previous section. It can also be seen that the fluorescence intensity of 1X is only about half of that of stock #3, despite the fact that the same amount of dye was used for the two batches. The personnel from Bangslabs, Inc. stated that this discrepancy could be due to the slight aging of the solvent, as the previous (stocks #1 to #6) and new (stocks 1X, 2X etc) batches were manufactured about 3 months apart. Due to the low intensity, stock 1/50X was not further considered as a suitable surrogate. Experimental results with the other batches will be discussed in the following section.

**Table 3.2: Average fluorescence intensities of the Rhodamine WT microsphere stocks which fluorescence distributions are shown in Figure 3.13**

| Stock | Fluorescence Intensity |
|---|---|
| #3 | 763 |
| 1/50X | 68.2 |
| 1/10X | 335 |
| 1X | 430 |
| 2X | 267 |
| 10X | 161 |
| 50X | 203 |

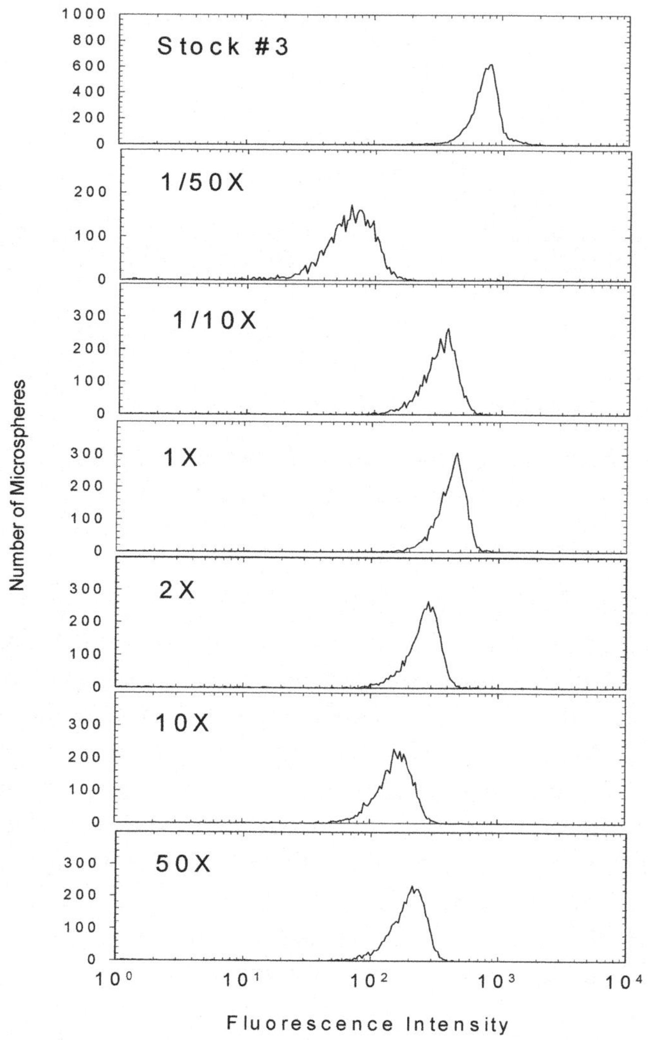


**Figure 3.13: Fluorescence intensity distributions obtained by flow cytometry analysis of different Rhodamine WT microsphere stocks within Gate R1 as shown in Figure 3.11**

### 3.2.2 Semi-batch Ozone Experiments with Microspheres (New Stocks)

The effect of ozone exposure on fluorescence intensity distribution of single microspheres from stocks 1/10X, 1X, 2X, 10X, 50X are shown in Figures 3.14 (a), (b), (c), and Figures 3.15 (a), (b), respectively. The control samples for the stocks had a single peak with relatively high fluorescence intensity. Increasing exposure to ozone resulted in a decrease in the average intensity, and led to the appearance of a second peak with much lower fluorescence.

The average fluorescence intensities in Figures 3.14 and 3.15 were used to calculate the normalized ratios ($I/I_o$) at different CT's, and the results are presented in Figure 3.16. The average intensity values and corresponding ratios are shown in Table 3.3. The kinetics of stock #3 is also included in the figure for comparison. As depicted in the figure, the decay kinetics for stocks 1X, 2X and 10X matched with that for stock #3 reasonably well. Stocks 1/10X and 50X, on the other hand, had a much faster kinetics. No explanations could be given for this observation.

Base on these experimental results, stocks 1/10X and 50X were dropped from further consideration. It was unexpected that stock 1X had lower average fluorescence intensity than that of stock #3, since the same amount of dye was used for both manufacturing processes. This finding revealed that the fluorescence intensity of the microspheres was very sensitive to slight variations in the manufacturing processes. Although microspheres from stocks 1X, 2X and 10X could serve as suitable surrogates because of their similar fluorescence decay kinetics to that of stock #3, stock 1X was selected because it had the highest fluorescence intensity. The higher the intensity, the higher ozone exposure the microspheres could encounter before the fluorescence intensity signal would disappear. Based on the results, an order was placed to produce 10 g of microspheres according to the protocol used for stock #3 (stock 1X) for use in full scale testing.

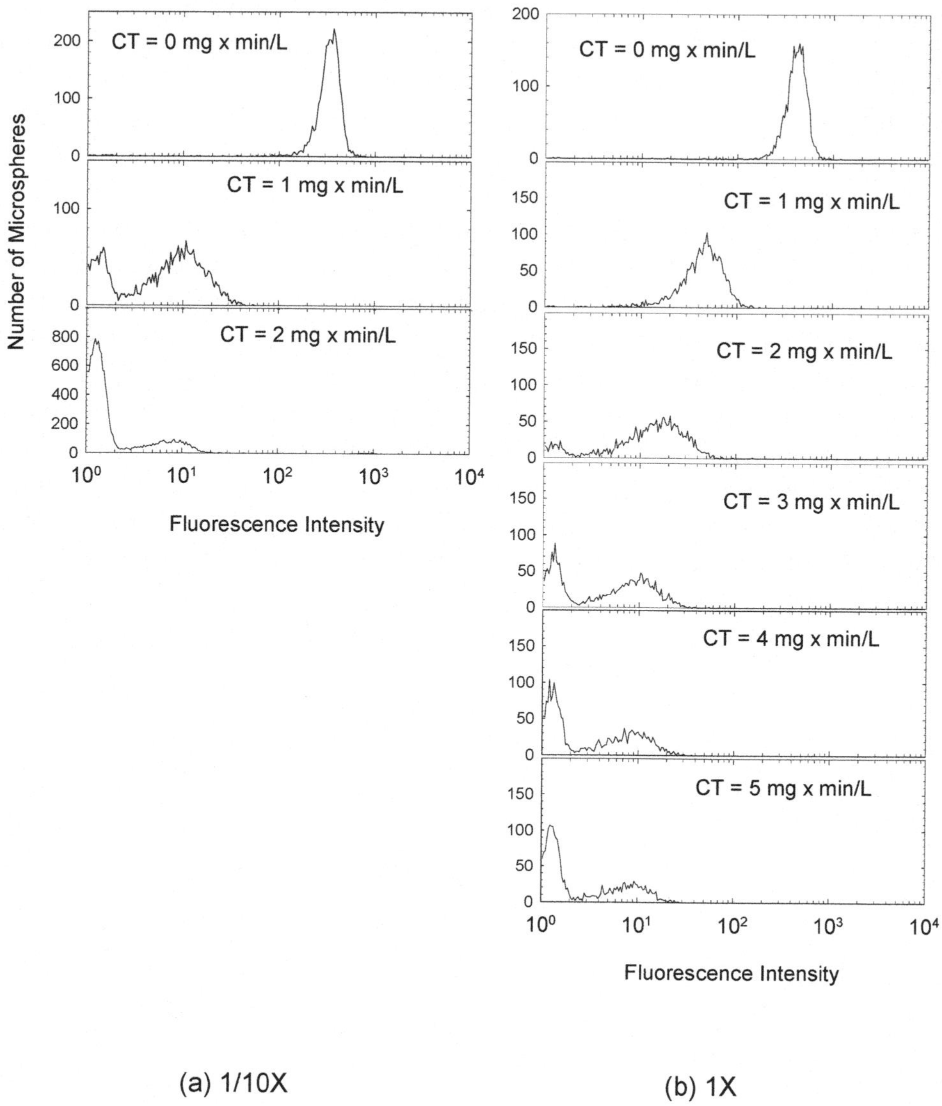


**Figure 3.14: Fluorescence intensity distribution obtained by flow cytometry analysis of Rhodamine WT microsphere stocks (a) 1/10X and (b) 1X**

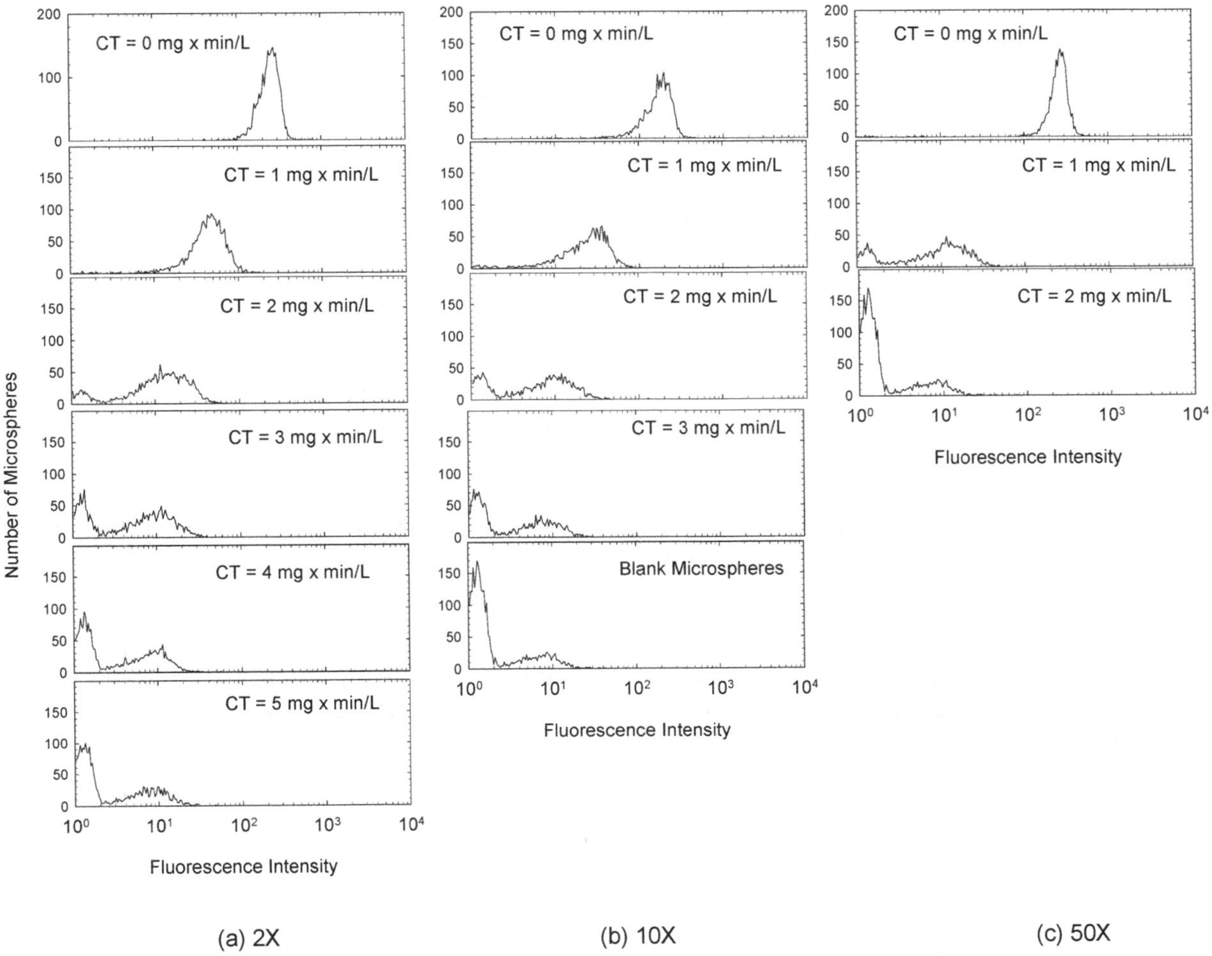


**Figure 3.15: Fluorescence intensity distribution obtained by flow cytometry analysis of Rhodamine WT microsphere stocks (a) 2X and (b) 10X and (c) 50 X**

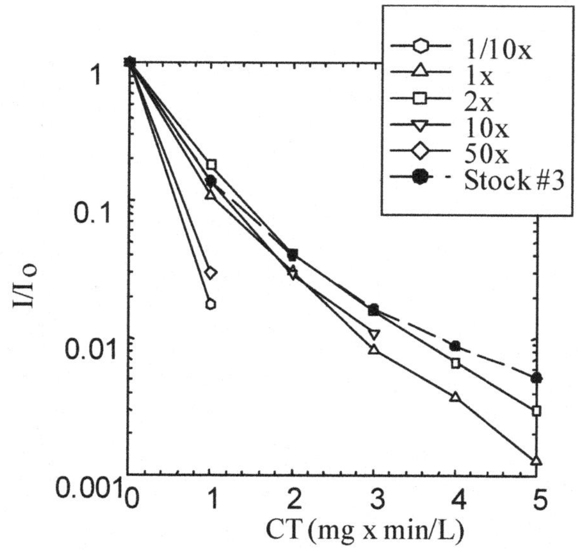


**Figure 3.16: Normalized fluorescence intensity decay of Rhodamine WT microsphere stocks with ozone exposure**

**Table 3.3: Average fluorescence intensities and corresponding normalized ratios of microspheres from stocks 1/10X, 1X, 2X, 10X and 50X**

| | Stock # | | | | | | | | | |
|---|---|---|---|---|---|---|---|---|---|---|
| $C \times T$ | 1/10X | | 1X | | 2X | | 10X | | 50X | |
| (mg×min/L) | $I$ or $I_0$ | $I/I_0$ | $I$ or $I_0$ | $I/I_0$ | $I$ or $I_0$ | $I/I_0$ | $I$ or $I_0$ | $I/I_0$ | $I$ or $I_0$ | $I/I_0$ |
| 0 | 331 | 1 | 408 | 1 | 244 | 1 | 175 | 1 | 260 | 1 |
| 1 | 5.78 | .0175 | 44.1 | 0.108 | 43.8 | .179 | 23.8 | .136 | 7.69 | 0.0300 |
| 2 | - | - | 12.4 | .0305 | 10.0 | .0410 | 5.12 | .0292 | - | - |
| 3 | - | - | 3.37 | .00827 | 3.85 | .0158 | 1.92 | .0109 | - | - |
| 4 | - | - | 1.51 | .00371 | 1.62 | .00663 | - | - | - | - |
| 5 | - | - | 0.52 | .00128 | 0.73 | .00300 | - | - | - | - |

# CHAPTER 4. EXPERIMENTAL STUDIES WITH YG MICROSPHERES

Experiments were performed with YG microspheres for both phases of full-scale testing. These microspheres were specifically chosen because they were used successfully in previous full-scale experiments (Mariñas *et al.*, 1999). This section presents the results from both bench-scale and full-scale testing, and a comparison to predictions made with the mathematical model described in Chapter 2.

## 4.1 PHASE I TESTING WITH YG MICROSPHERES

### 4.1.1 Semi-batch Ozonation Experiments

The objectives of the bench scale experiments were to optimize the microspheres in terms of fluorescence decay kinetics, and to develop a correlation between the fluorescence decay kinetics of microspheres and the inactivation kinetics of *C. parvum* oocysts. The fluorescence intensity distributions of preozonated microspheres exposed to ozone at various *CT* values are shown in Figure 4.1. As shown in the figure, the overall fluorescence decreased with ozone exposure, due to the reaction of the dye with ozone.

The curves in Figure 4.1 were correlated with the inactivation kinetics of *C. parvum* oocysts by the threshold concept (Chiou *et al.*, 1997). Microspheres with fluorescence intensity above and below a certain fluorescence intensity threshold were considered equivalent to viable and non-viable oocysts, respectively. The resulting microsphere "inactivation" curves are shown in Figure 4.2 for fluorescence intensity thresholds values in the range of 26 to 29 at an interval of 1. Two sets of *C. parvum* inactivation kinetics at the full-scale experimental temperature of 22.3°C were also included in the figure. The solid line corresponded to the inactivation kinetics reported by Rennecker *et al.* (1999), and the dashed line was obtained based on the *CT* equation developed by Clark *et al.* (2002). The major difference between the two sets of kinetics was the occurrence of a lag-phase (shoulder) region for data provided by Rennecker *et al.* (1999), during which little or no inactivation occurs. This characteristic is not shown in the inactivation kinetics proposed by Clark *et al.* (2002) because these authors omitted the lag phase and correlated several data sets obtained from the literature with the pseudo-first order Chick-Watson model. The inactivation kinetics based on Rennecker *et al.* (1999) was chosen for Phase I of full-scale at the time of the experiments, the *CT* equation proposed by Clark *et al.* (2002) had not been published. Accordingly, the microspheres were designed to match the inactivation data by Rennecker *et al.* (1999). From Figure 4.2, it can be seen that a threshold value of 27 provided the best correlation between the microsphere fluorescence decay and oocyst inactivation kinetics. This threshold value will be used to estimate the disinfection efficiency along the full-scale ozone contactor during Phase I testing as described subsequently.

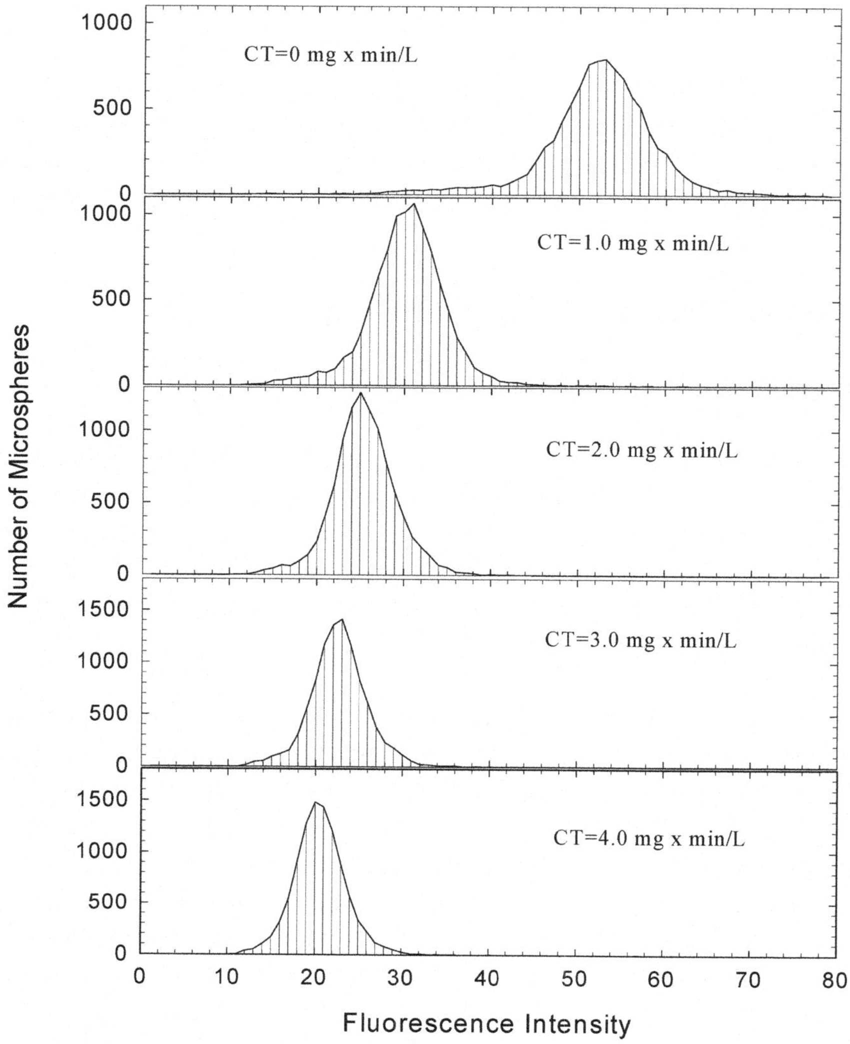


**Figure 4.1: Fluorescence Intensity of YG microspheres after ozone exposures in the range of 0 to 4 mg×min/L**

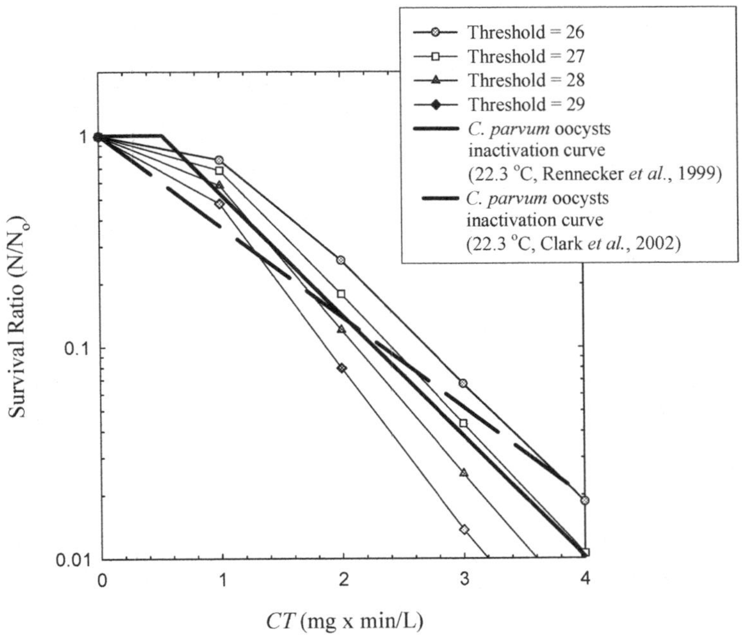


**Figure 4.2. Comparison of ozone "inactivation" kinetics for YG microspheres at threshold values in the range of 26 to 29, and *C. parvum* oocysts**

### 4.1.2 Batch Ozonation Experiments

Batch ozonation experiments were performed to determine the ozone decomposition kinetics in Phase I raw water. The decomposition kinetics was investigated at various initial dissolved ozone concentrations. The results, shown in Figure 4.3, revealed that the ozone decomposition curve was characterized by the occurrence of a relatively fast initial decay followed by slower first-order decomposition. These results also showed that the decomposition kinetics was dependent on the initial ozone concentration, with the rate of decay increasing with decreasing initial dosage. The formation kinetics of bromate was not characterized because the quenching agent used during Phase I of full-scale testing interfered with the bromate analyses and consequently no data could be obtained. The quenching was changed for Phase II as discussed in Chapter 2 in order to avoid this problem.

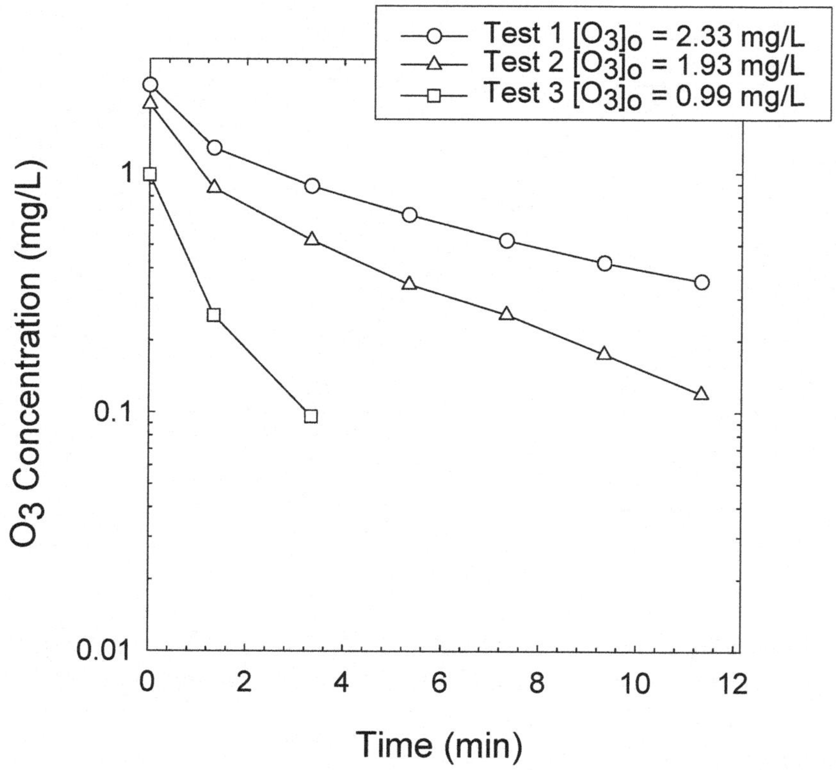


**Figure 4.3: Decomposition of ozone in raw water entering the ACWD contactor during Phase I of full-scale testing**

### 4.1.3 Full-Scale Demonstration Experiments

Representative results obtained from the flow cytometry analyses of samples taken for the six tests performed during Phase I of full-scale testing are shown in Figure 4.4. As depicted in the figure, background fluorescence peaks with relatively low intensity were observed for all samples, likely due to the presence of algae, bacteria and other fluorescent particles in the raw water (Mariñas *et al.*, 1999). As a result, the particles with low fluorescence (less than approximately 10 to 15 on the scale shown) were disregarded when estimating the $N/N_o$ ratio for microspheres. The threshold value of 27, obtained from the semi-batch experiments (see Figure 4.2) was used to determine the corresponding "survival" ratio of microspheres, which was then used to assess the viability of *C. parvum* oocysts.

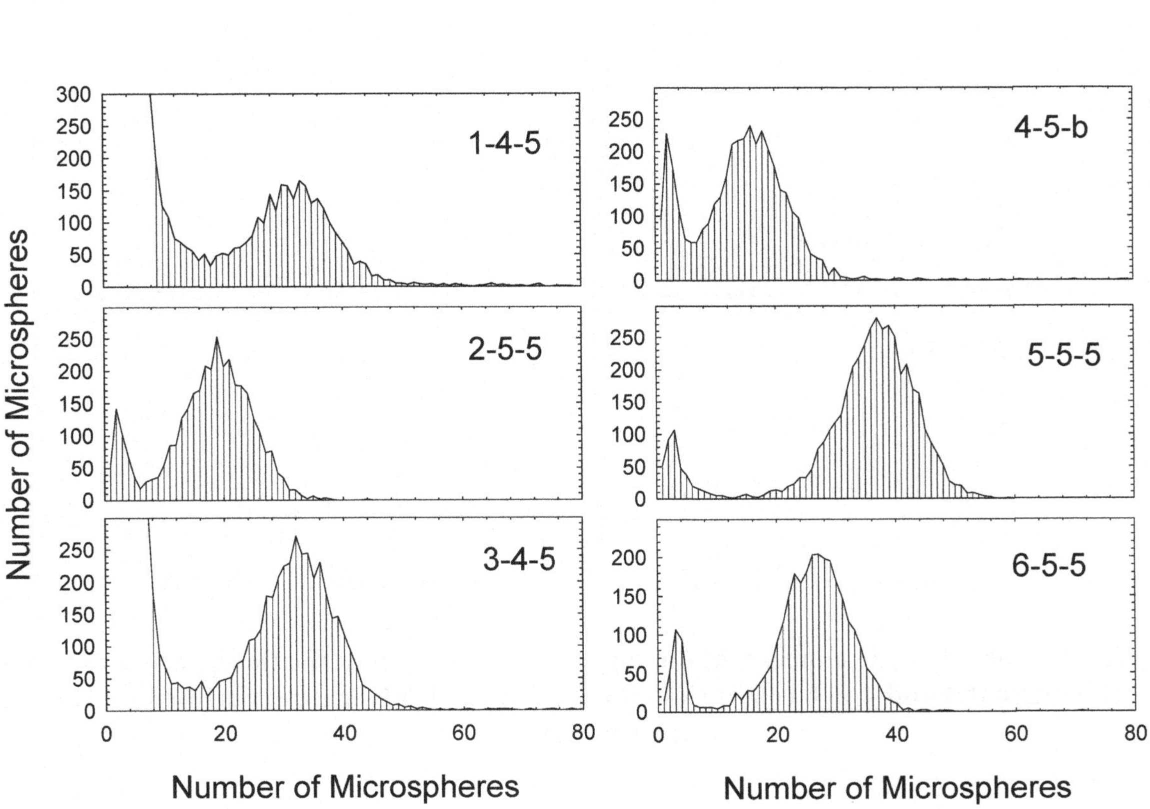


**Figure 4.4: Flow cytometry analysis of selected samples obtained from Phase I of full-scale testing with YG microspheres (The samples are labeled as Test# - Chamber# - Port# consistent with Figure 2.2)**

Dissolved ozone concentrations and *C. parvum* oocyst inactivation profiles throughout the contactor for the six tests performed during Phase I (see Table 2.2 for experimental conditions used) are shown in Figure 4.5. As depicted in the figure, the inactivation efficiency increased, as expected, with ozone exposure along the contactor for each test. Furthermore, as also expected, greater overall inactivation efficiency was achieved throughout the contactor at higher ozone dosage. The *CT* value corresponding to the overall microsphere inactivation efficiency achieved in the contactor for each test was then obtained graphically from Figure 4.2. Each *CT* value was obtained by introducing the corresponding $N/N_o$ value in the curve for the optimum fluorescence intensity threshold of 27 in Figure 4.2. The resulting "microsphere-based" *CT* values are compared to the corresponding *CT* values (using both $T_{10}$ and *HRT*) obtained from the dissolved ozone profiles in Table 4.1. As shown in the table, the microsphere-based values were consistently lower than the ozone profile-based values. This discrepancy will be further discussed in Section 4.1.4.

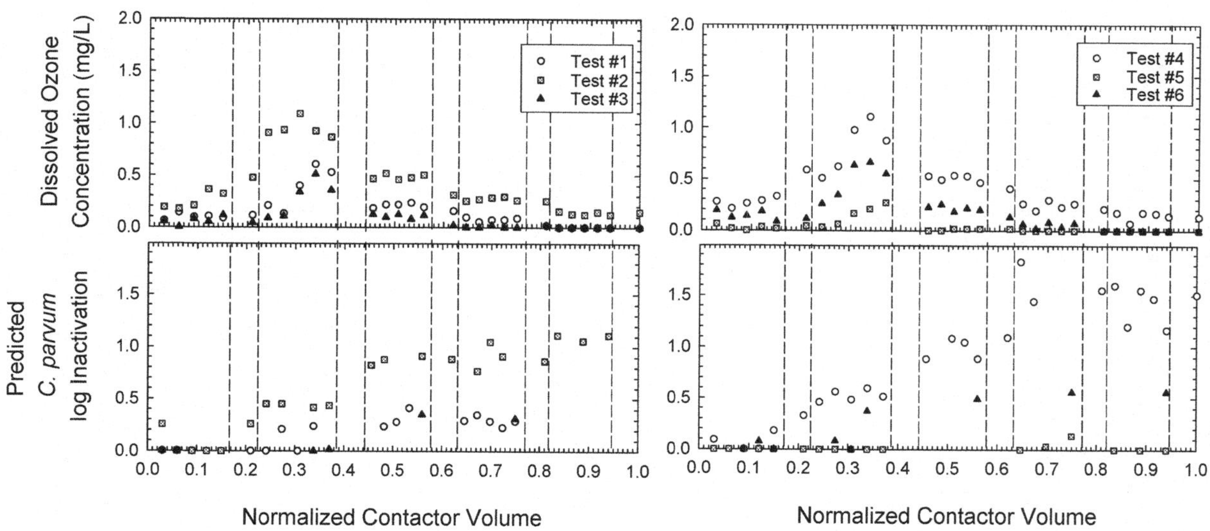


**Figure 4.5: Dissolved ozone concentration, and *C. parvum* log inactivation profiles for tests 1 to 3 (left side plots) and 4 to 6 (right side plots) of Phase I full-scale experiments with YG microspheres**

**Table 4.1: Comparison of *CT* values based on overall microsphere inactivation data and dissolved ozone concentration profiles (Phase I)**

| | | | Ozone Conc. Profile-Based | | "Microsphere-Based" |
|---|---|---|---|---|---|
| Test No. | Water Flow Rate (MGD) | Ozone Dosage (mg/L) | $CT_{10}$ (mg×min/L) | $C \times HRT$ (mg×min/L) | $CT$ (mg×min/L) |
| 1 | 11.0 | 3.58 | 1.56 | 2.22 | 1.13 |
| 2 | 11.1 | 4.79 | 4.20 | 6.00 | 2.63 |
| 3 | 11.3 | 3.00 | 0.87 | 1.24 | 1.13 |
| 4 | 9.4 | 4.79 | 5.27 | 7.52 | 2.79 |
| 5 | 9.4 | 2.68 | 0.42 | 0.604 | 0.25 |
| 6 | 9.4 | 3.66 | 2.15 | 3.07 | 1.36 |

### 4.1.4 Mathematical Modeling

The mathematical model developed by Kim *et al.* (2002a,b) presented in Section 2.5 was used to predict the ozone concentration and inactivation profiles throughout the ACWD contactor for the operating conditions investigated in Phase I of full-scale testing. As discussed in Section 4.1.3, the ozone decomposition kinetics varied significantly with the initial ozone dosage, with the apparent decay rate increasing with decreasing dosage. Figure 4.5 revealed that the ozone concentration within the contactor during the full-scale experiment was always below 1.5 mg/L. Accordingly, the kinetic parameters obtained for the batch experiment at the lowest ozone dosage of 0.99 mg/L (see Figure 4.3) were used for modeling purposes. The parameters obtained from fitting the ozone decay curve in Figure 4.3 were $D_o$ = 0.52 mg/L as $O_3$, and $k_s$ = 0.008 $s^{-1}$.

The data obtained for a tracer test performed as part of a previous characterization of the ACWD contactor hydrodynamics was also analyzed in order to obtain an appropriate dispersion number (*d*). The data obtained for a tracer test monitored at the outlet of the downflow section of the first ozone chamber is shown in Figure 4.6. The corresponding contactor operating conditions were a gas flow rate of 105 scfm, and a water flow rate of 14.5 MGD. A least square method was used to obtain the dispersion number value $d = 3$. The tracer curve corresponding to this dispersion number, also shown in Figure 4.6, revealed that the axial dispersion model provided a good representation of the chamber hydrodynamics. Although the experimental conditions used during Phase I, as well as during Phase II, of full-scale testing were somewhat different, the dispersion number $d = 3$ was used in all predictions. This was found to be an adequate approach based on the fact that, as depicted in Figure 4.6 and supported by the high *d* value, the hydrodynamics of the down-flow chamber approached that of a continuous flow stirred tank reactor (CSTR).

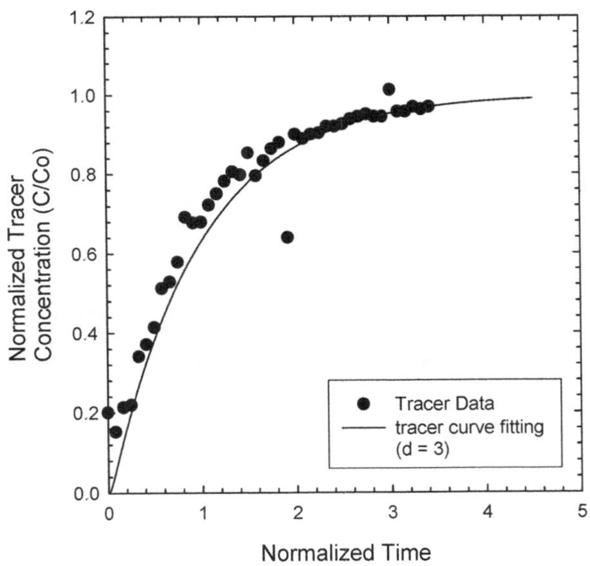


**Figure 4.6: Tracer test performed on the down-flow portion of the first chamber of the ACWD ozone contactor (gas flow rate = 105 scfm; water flow rate = 14.5 MGD)**

The ozone concentration and *C. parvum* inactivation profiles predicted with the model described in Chapter 2 after incorporating ozone decay parameters and dispersion number, as well as the *C. parvum* inactivation kinetics from Rennecker *et al.* (1999) are shown in Figure 4.7 (Tests 1-3) and Figure 4.8 (Tests 4 to 6). As depicted in the figures, the model over-predicted the level of *C.*

*parvum* inactivation. Although no clear explanation could be provided for such discrepancy some factors potentially contributing to these effects were identified. First, the water quality of the contactor influent sample used to obtain the ozone decomposition kinetics might have changed between the time the sample was taken during the full-scale experiments and that when the final batch ozonation experiments were performed at the University of Illinois. This would have led to an underestimation of the ozone decay kinetics parameters, and corresponding over-prediction of dissolved ozone concentration and *C. parvum* inactivation. Secondly, scanning electron microscopy (SEM) analyses of the microsphere stock used in Phase I of full-scale testing revealed the occurrence of microsphere clumping as shown in Figure 4.9(a). The clumping was probably due to the absence of surfactants in the microspheres suspension, which was removed to provide a stock as clean as possible. Furthermore, the samples prepared for SEM analysis consisted of microspheres at a concentration of about $10^{-8}$ g/L, where the microspheres stock used in the full-scale tests had a concentration of about 0.3 g/L. Therefore, the clumping effect was likely more severe in the full-scale experiments compared to that shown in Figure 4.9(a). Clumping would decrease the exposure of the microspheres to ozone, which would result in a slower fluorescence decay, and thus in under-predicted inactivation. As a result, the data obtained from Phase I of full-scale testing were not considered representative to the actual performance of the contactor, but provided useful preliminary information to ensure better success during Phase II of full-scale testing.

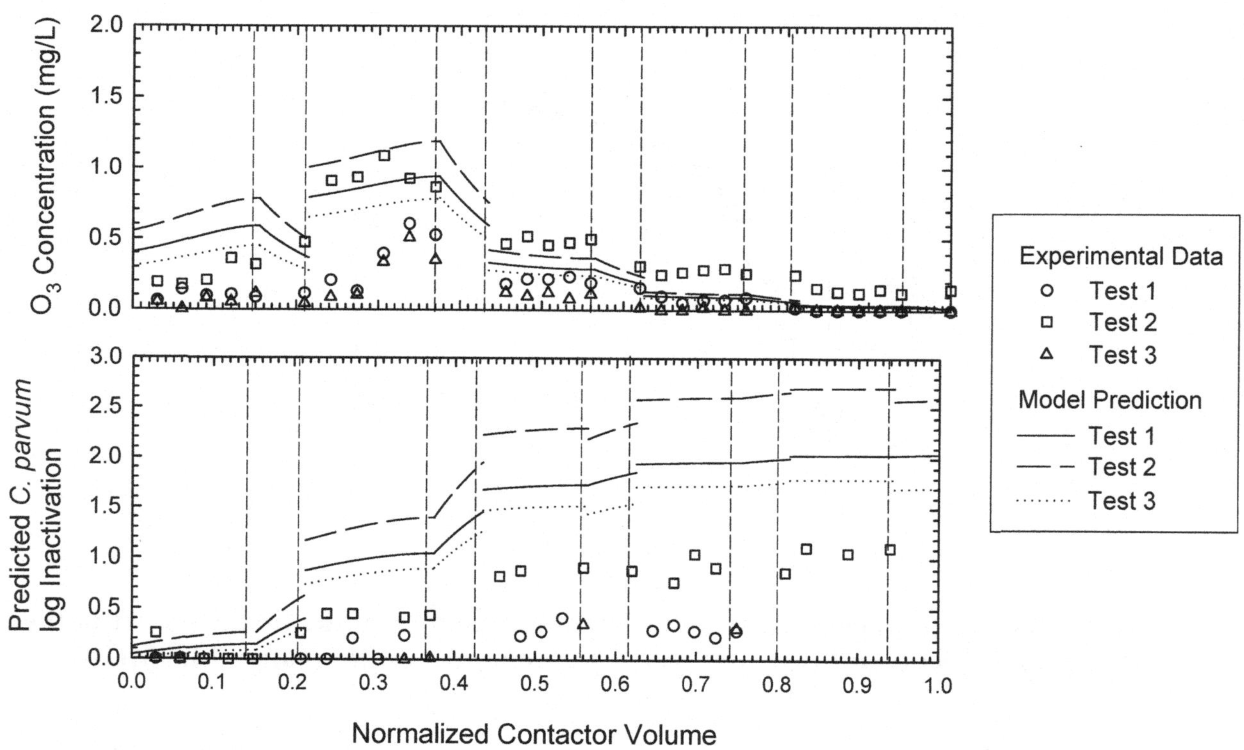


**Figure 4.7: Predicted ozone concentration and *C. parvum* inactivation profiles for tests 1 to 3 part of Phase I full-scale testing with YG microspheres**

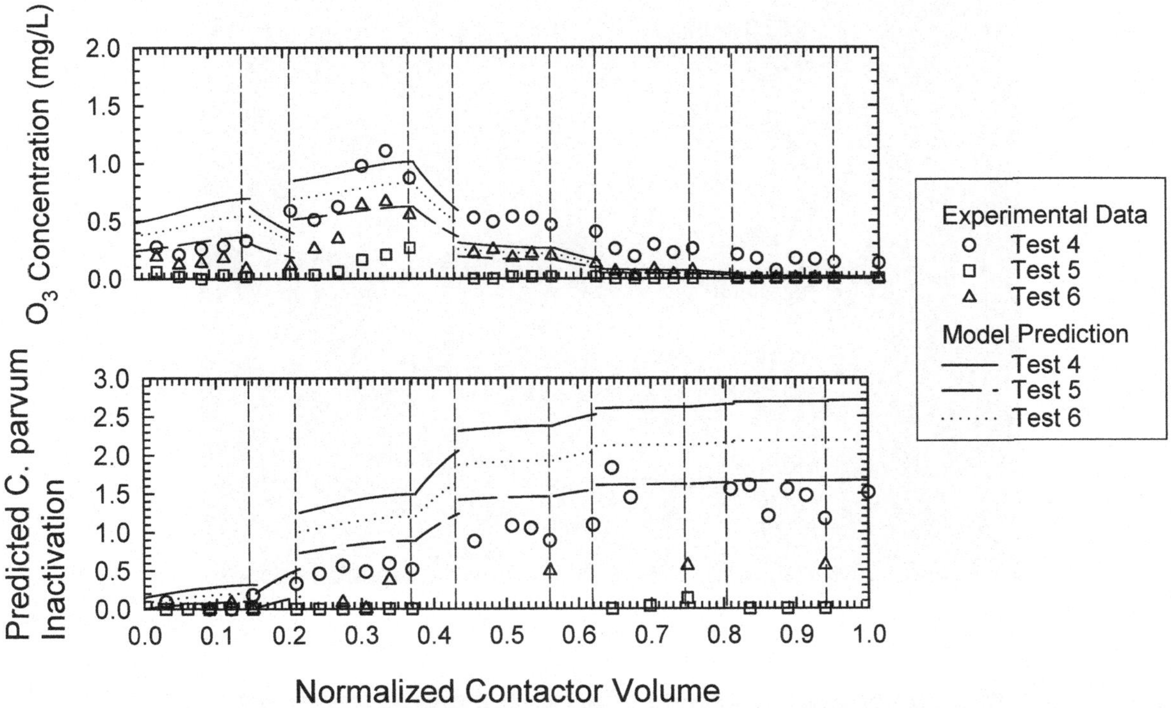


**Figure 4.8: Predicted ozone concentration and *C. parvum* inactivation profiles for tests 4 to 6 part of Phase I full-scale testing with YG microspheres**

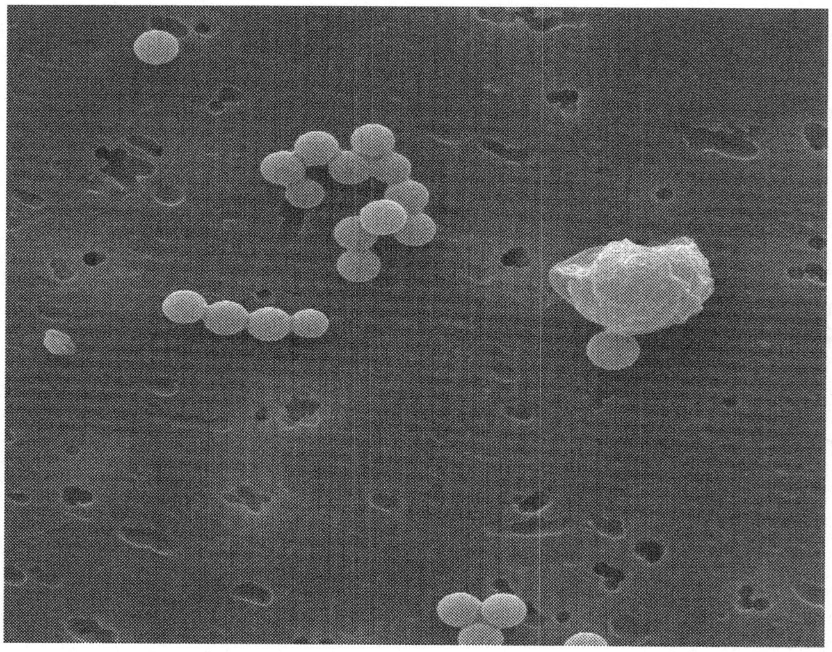

(a)

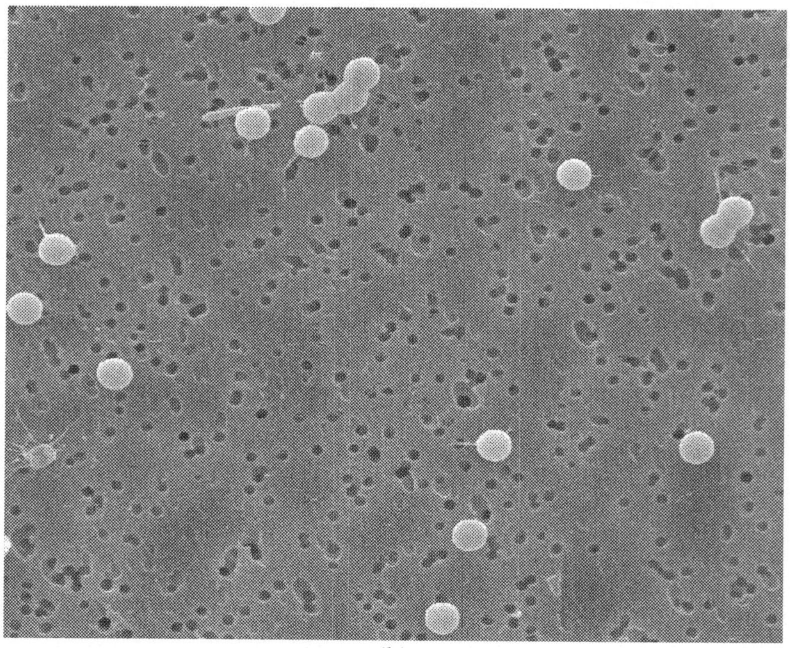

(b)

**Figure 4.9: Scanning electron micrograph of YG microspheres for (a) Phase I and (b) Phase II of full scale experiments**

## 4.2 PHASE II EXPERIMENTS WITH YG MICROSPHERES

### 4.2.1 Semi-batch Ozonation Experiments

The objectives of semi-batch ozone experiments were to develop a correlation between the fluorescence decay kinetics of microspheres and the inactivation kinetics of *C. parvum* oocysts. Fluorescence intensity distributions of microspheres exposed to ozone *CT* levels of 0 to 5 mg×min/L are shown in Figure 4.10. Following the same approach used for semi-batch curves for Phase I of testing, the curves in Figure 4.10 were correlated with the *C. parvum* oocysts disinfection kinetics using the threshold concept (Chiou *et al.*, 1997). The resulting microsphere "inactivation" curves for fluorescence intensity thresholds in the range of 41 to 44, at an interval of 1, are shown in Figure 4.11. The predicted *C. parvum* oocyst inactivation kinetics (Clark *et al.,* 2002) at 19.1 $^{o}$C, the raw water temperature measured during test 1 of Phase II full-scale testing, is also shown in Figure 4.11. As depicted in the figure a threshold value of 43 provided the best correlation between the microsphere fluorescence decay and oocyst inactivation kinetics for ozone exposures in the range of 0 to 5 mg×min/L. This threshold value will be used to estimate the disinfection efficiency along the full-scale ozone contactor as described subsequently.

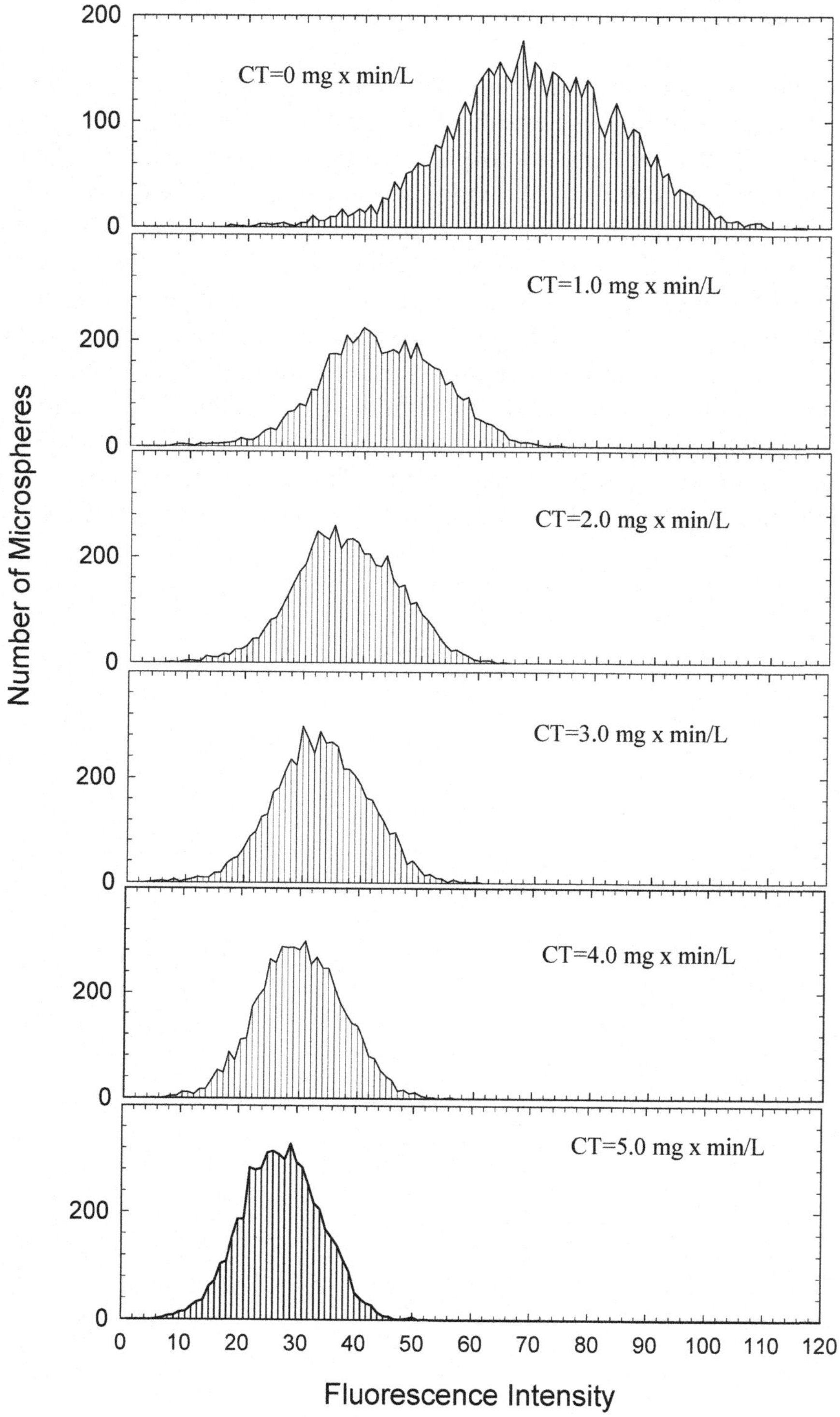


**Figure 4.10: Fluorescence intensity of YG microspheres after ozone exposures of 0 to 5.0 mg×min/L**

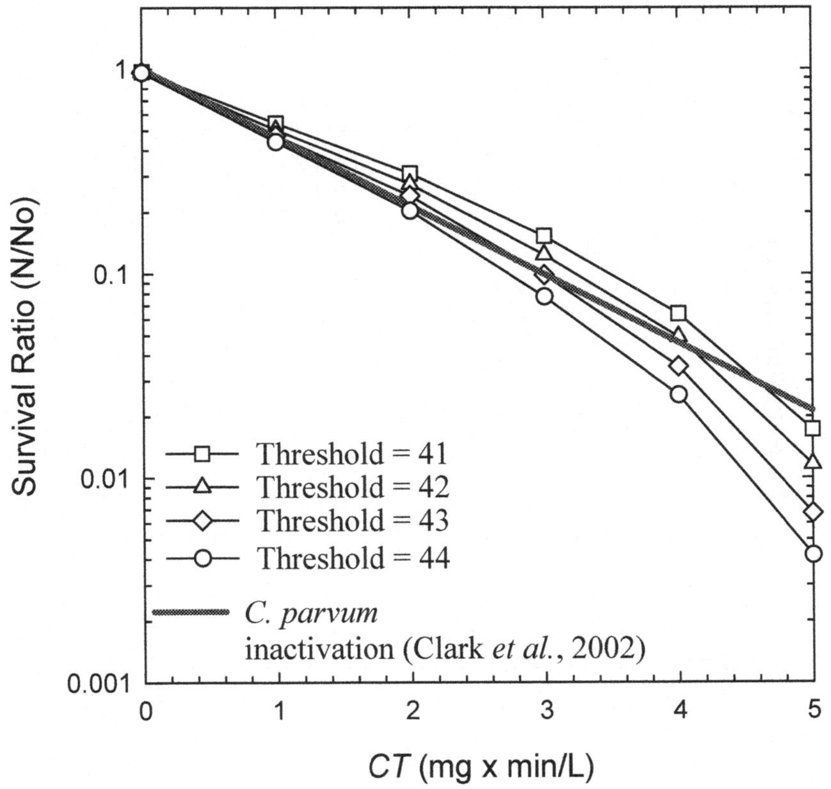


**Figure 4.11: "Inactivation" kinetics of YG microspheres for threshold values of 41 to 44, and *C. parvum* oocysts**

### 4.2.2 Batch Ozonation Experiments

Experiments were performed with the batch ozonation setup in order to determine the ozone decomposition kinetics and bromate formation in the Phase II raw water at 20 °C. The decomposition kinetics of ozone at various initial dosages is presented in Figure 4.12. Similar to the observation for Phase I water, the ozone decomposition kinetics was characterized by an initial fast decay followed by a first order decrease in concentration. The data also showed that the decomposition kinetics were once again dependent on the initial ozone concentration. As also depicted in the figure, and in agreement with the data from Phase I testing, the ozone decay rate was relatively faster at lower ozone dosage.

Bromate samples were taken at the end of ozone decay experiment. The resulting bromate concentrations were then plotted against the corresponding *CT* (Figure 4.13) obtained by integrating the ozone decay curve in Figure 4.12. As depicted in the figure, the formation of bromate in the ACWD water is linear with respect to ozone exposure, regardless of the initial ozone dosage used. The results showed that for every unit of ozone exposure (mg×min/L), 5.63

μg/L of bromate was produced. This linear relationship between ozone exposure and bromate concentration is used in the mathematical model to predict the concentration of bromate for Phase II of full-scale testing as described subsequently.

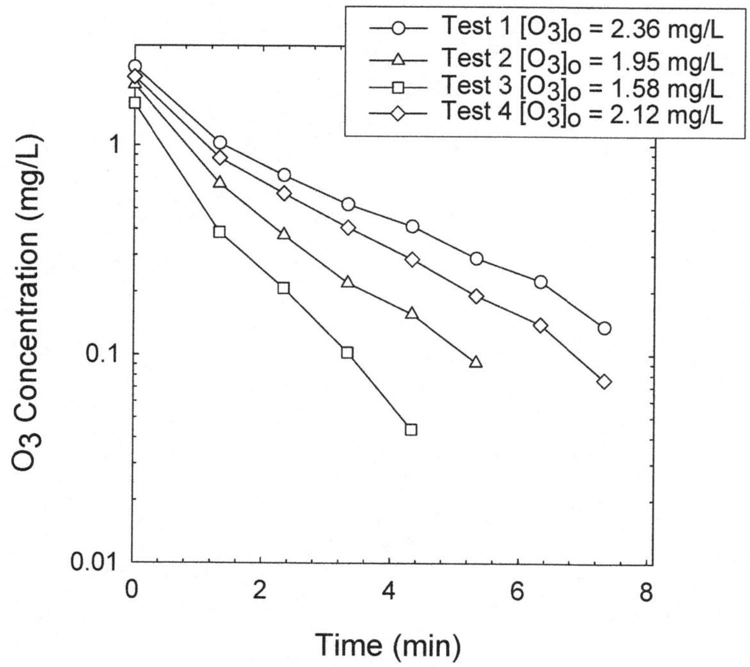


**Figure 4.12: Ozone decomposition curves in ACWD raw water from Phase II of full-scale testing**

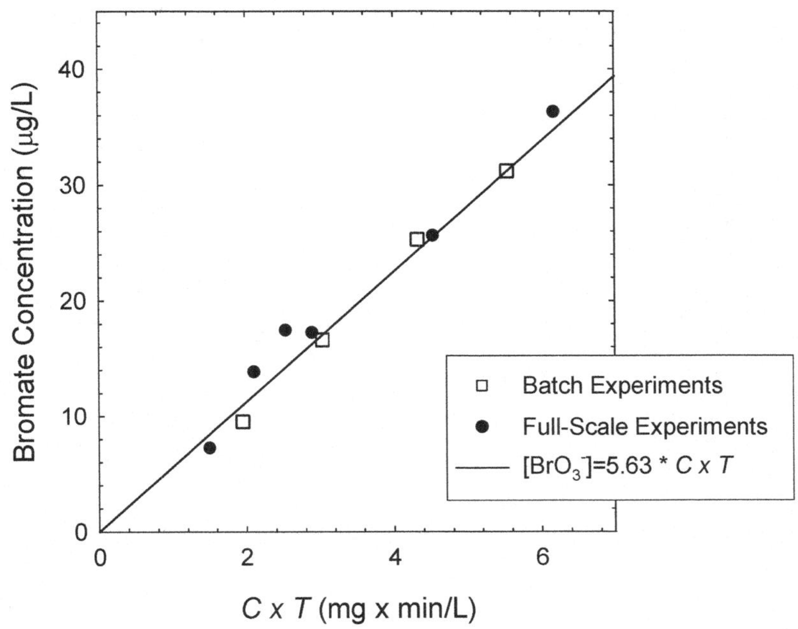


**Figure 4.13: Bromate concentration of samples taken from batch ozonation tests and the effluents of full-scale experiments**

### 4.2.3 Full-Scale Demonstration Experiments

Representative results obtained from the flow cytometry analyses of samples taken for each of the six experiments performed during Phase II of full-scale testing are shown in Figure 4.14. The data correspond to samples taken from the final contactor effluent. Similar to the observation during Phase I, background fluorescence with relatively low intensity was observed in the raw water, probably due to algae, bacteria and other fluorescent particles present in the source water. As a result, the particles within the peak of low fluorescence intensity (less than about 10 on the scale shown) were disregarded when estimating the corresponding $N/N_o$ values. Based on the correlation shown in Figure 4.11, the threshold value of 43 was used to determine the corresponding "survival" ratio of microspheres, which was then used to assess the viability of *C. parvum* oocysts.

The dissolved ozone concentration, *C. parvum* oocysts inactivation, and bromate formation profiles within the contactor for all six experiments performed during Phase II of full-scale testing are shown in Figure 4.15. As depicted in the figure, increasing the ozone dosage resulted in an increase in dissolved ozone concentration, and corresponding greater inactivation efficiency and bromate formation. The resulting "microsphere-based" *CT* values are compared to the corresponding *CT* values (using both $T_{10}$ and *HRT*) obtained from the dissolved ozone profiles in Table 4.2. It is interesting that contrary to Phase I, the "microsphere-based" values were consistently higher than the ozone profile-based values. No explanation can be provided at this time, but it probably suggests that the full-scale protocol may need to be refined for better quality control.

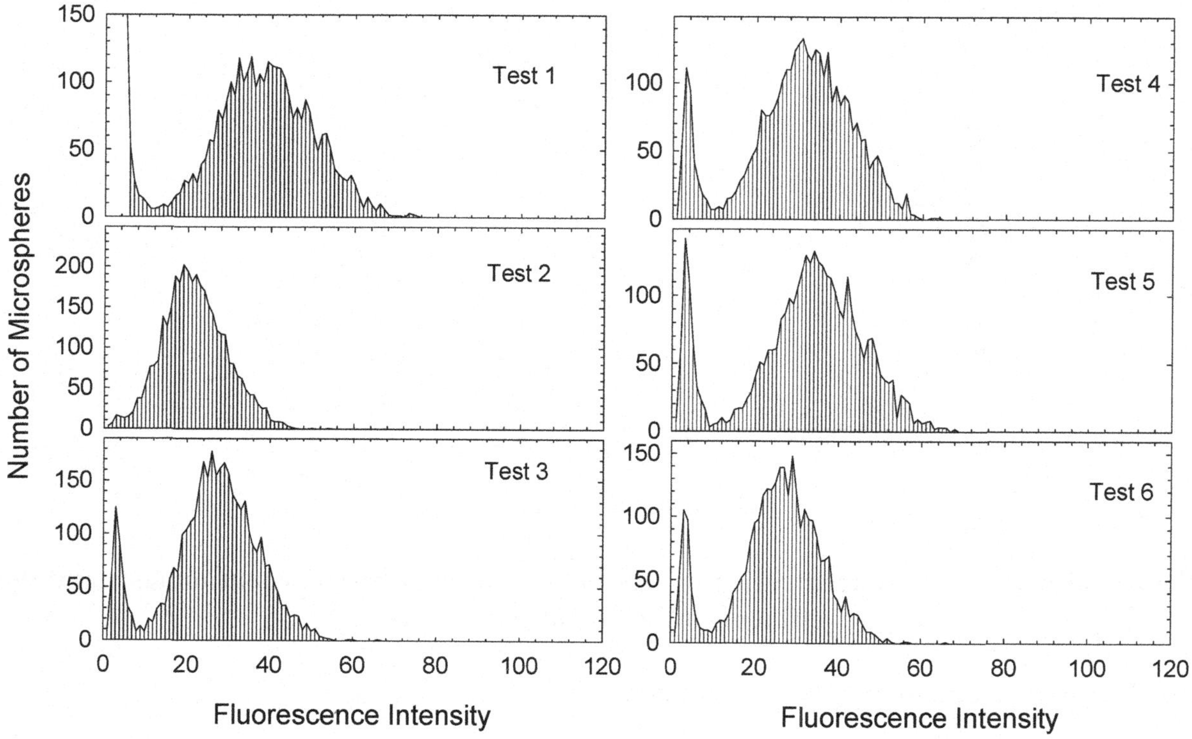


**Figure 4.14: Fluorescence intensity distribution of YG microspheres recovered from effluent samples of the six experiments performed during Phase II of full-scale testing**

**Table 4.2: Comparison of *CT* values based on overall microsphere inactivation data and dissolved ozone concentration profiles (Phase II)**

| | | | Ozone Conc. Profile-Based | | "Microsphere-Based" |
|---|---|---|---|---|---|
| Test No. | Water Flow Rate (MGD) | Ozone Dosage (mg/L) | $CT_{10}$ (mg×min/L) | $C \times HRT$ (mg×min/L) | $CT$ (mg×min/L) |
| 1 | 9.5 | 2.65 | 0.907 | 1.30 | 1.55 |
| 2 | 9.7 | 4.20 | 3.73 | 5.33 | 5.64 |
| 3 | 9.5 | 3.29 | 1.75 | 2.49 | 3.41 |
| 4 | 15.5 | 3.19 | 1.53 | 2.18 | 2.56 |
| 5 | 15.5 | 2.83 | 1.27 | 1.81 | 2.10 |
| 6 | 15.5 | 4.03 | 2.74 | 3.91 | 4.15 |

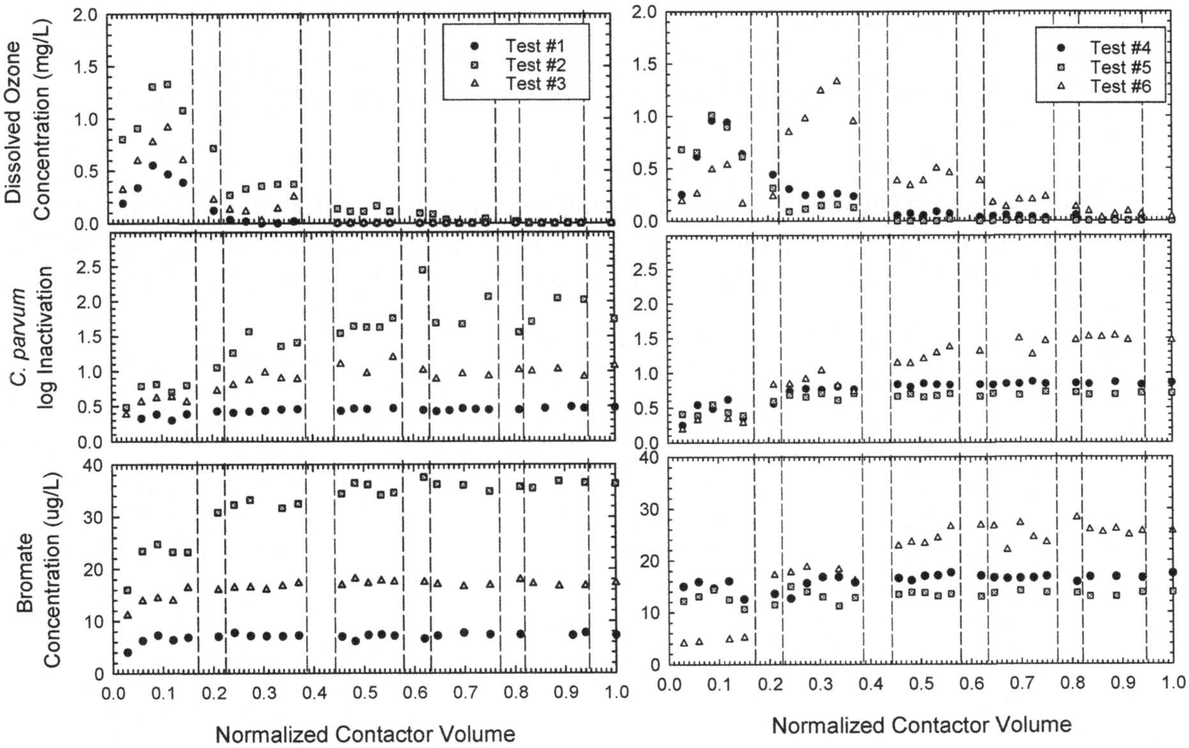


**Figure 4.15: Dissolved ozone concentration, *C. parvum* log inactivation and bromate concentration profiles for tests 1 to 3 (left plots) and 4 to 6 (right plots) obtained for Phase II of full-scale testing**

### 4.2.4 Mathematical Modeling

The mathematical model presented in Chapter 2 was used to predict the ozone concentration, *C. parvum* inactivation, and bromate formation profiles of the contactor under the operating conditions used during Phase II of full-scale testing (see Table 2.3). The ozone decomposition data sets in Figure 4.12 were fitted to obtain the corresponding parameters $D_o$ and $k_s$ at each ozone dose. Similar to the observation during Phase I, it was found that a single set of parameters could not represent all curves. However, because the dissolved ozone concentrations throughout the ozone full-scale contactor were below 1.5 mg/L for tests 1, and 3-5 it was decided to use the parameters corresponding to the lowest ozone dose of 1.59 mg/L investigated in batch experiments. The corresponding parameters were $D_o$ = 0.71 mg/L as $O_3$ and $k_s$ = 0.011 $s^{-1}$. Because of the higher concentrations reached for tests 2 and 6, the kinetic parameters $D_o$ = 0.85 mg/L as $O_3$ and $k_s$ = 0.0080 $s^{-1}$, based on the batch test with initial dosage of 1.98 mg/L, were used. The correlation between ozone exposure and bromate formation, as obtained from Figure 4.13, was used in the model prediction for bromate concentration. Kim (2002) has shown that bromate formation is independent of hydrodynamics because of the low conversion for this reaction. The full-scale bromate data was in agreement with this previous observation as shown in Figure 4.13. As depicted in the figure, a plot of the bromate concentrations for the six Phase II full-scale tests against the corresponding *CT* values (calculated based on dissolved ozone concentration profiles and *HRT*) was in excellent agreement with the batch data. This consistency took place despite the fact that the hydrodynamic conditions in the full-scale reactor deviated significantly from ideal plug flow reactor (PFR) conditions. Therefore, for simplicity, the down-flow and up-flow sections of the contactor were modeled with CSTR and PFR models, respectively.

The model prediction of the ozone concentration, *C. parvum* inactivation, and bromate formation profiles are shown in Figures 4.16 for tests 1, 2 and 3, and Figure 4.17 for tests 4, 5 and 6. As shown in the figure, the model prediction matches reasonably well for all of the experiments.

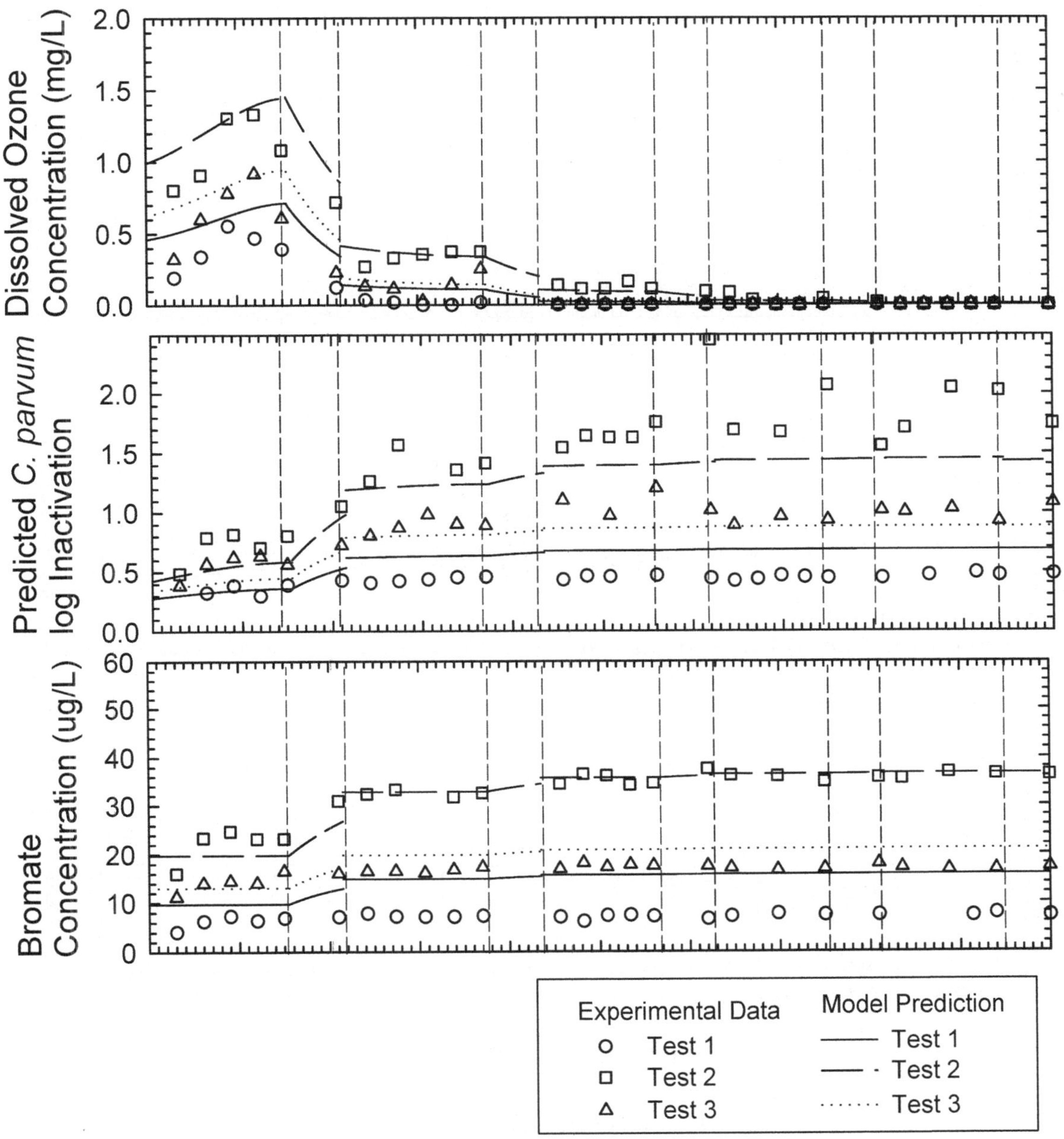


**Figure 4.16: Model prediction of dissolved ozone concentration, *C. parvum* inactivation, and bromate concentration profiles for tests 1 to 3 of Phase II experiments**

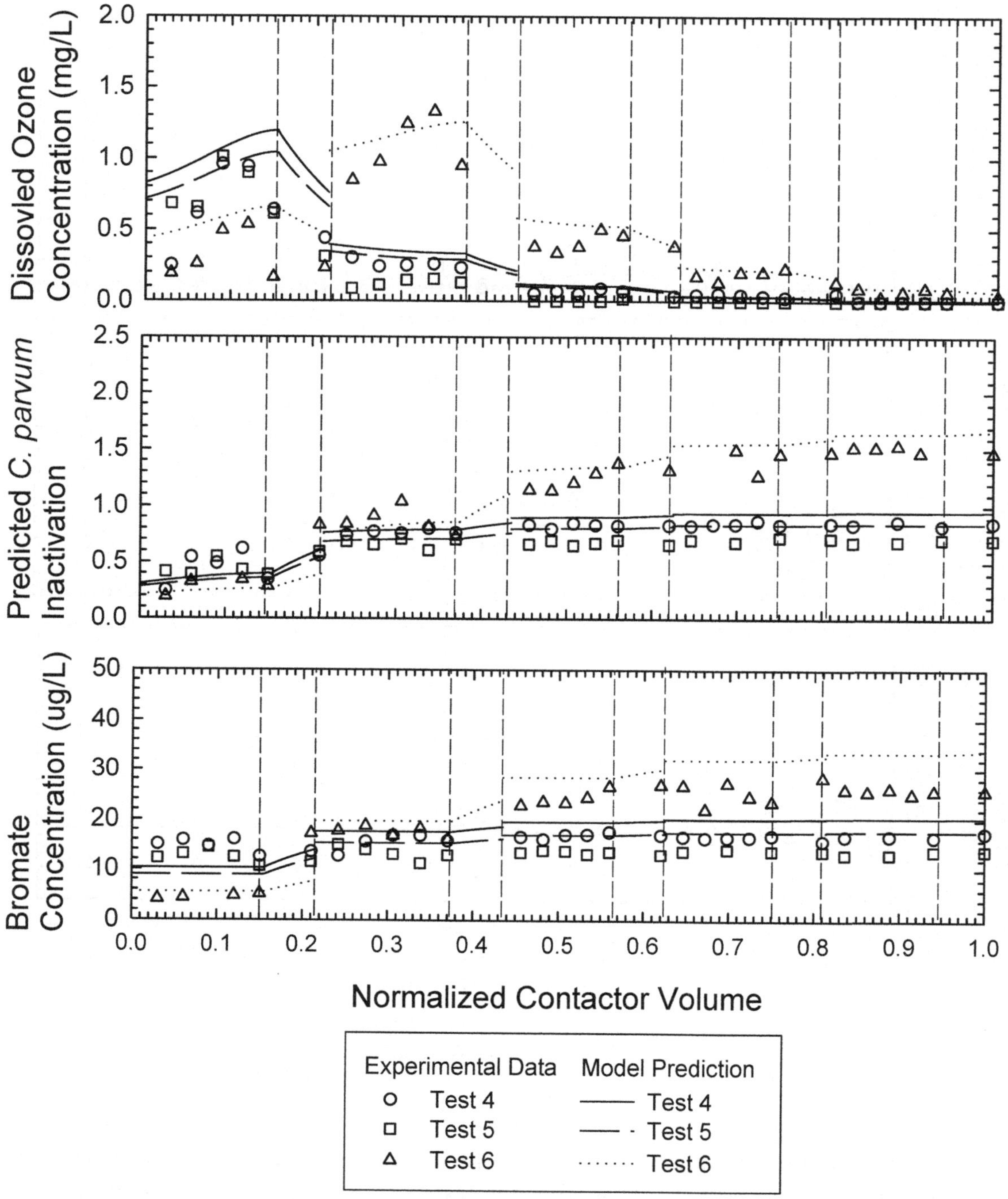


**Figure 4.17: Model prediction of dissolved ozone concentration, *C. parvum* inactivation, and bromate concentration profiles for tests 4 to 6 of Phase II experiments**

## 4.3 EFFECT OF REACTOR DESIGN ON REACTOR PERFORMANCE

The results shown in Figures 4.16 and 4.17 and the corresponding predicted curves for Phase II of full-scale testing revealed that the bromate MCL would be exceeded when applying conditions resulting in *C. parvum* inactivation levels in excess of 0.5 log. Although an approach to lower the levels of bromate formation is to lower the pH of the feed water, such goal could also be achieved by identifying and changing reactor design components and configuration conditions responsible for the relatively high level of back-mixing revealed by tracer tests. Although identifying such components and conditions are beyond the scope of this study, the model could be applied to provide information on the minimum level of bromate that could be formed while achieving various levels of inactivation efficiency.

Model predictions were made for the extreme case in which each chamber of the ACWD contactor would approach the hydrodynamic conditions of an ideal PFR. The simulations were performed with the goal of obtaining the same level of inactivation as those shown in Figures 4.16 and 4.17. In contrast to bromate formation, *C. parvum* inactivation is a high conversion reaction (i.e., high level of inactivation), and therefore, the contactor hydrodynamics would affect the reaction efficiency. Accordingly, the ozone dose was lowered by trial and error until the same levels of inactivation as those predicted from the actual experimental conditions were achieved. The resulting simulated curves for both *C. parvum* inactivation and bromate formation under PFR conditions are compared in Figures 4.18 (Tests 1-3) and 4.19 (Tests 4-6) to those obtained for the actual hydrodynamic conditions occurring during Phase II testing. As depicted in the figure, decrease in bromate formation was attained by improving the hydrodynamics of the reactor, with the greatest improvement observed for the highest level of conversion of the *C. parvum* inactivation reaction. These results reveal that comprehensive strategies aimed at controlling bromate formation could include a combination of approaches including changing water quality and improving reactor design.

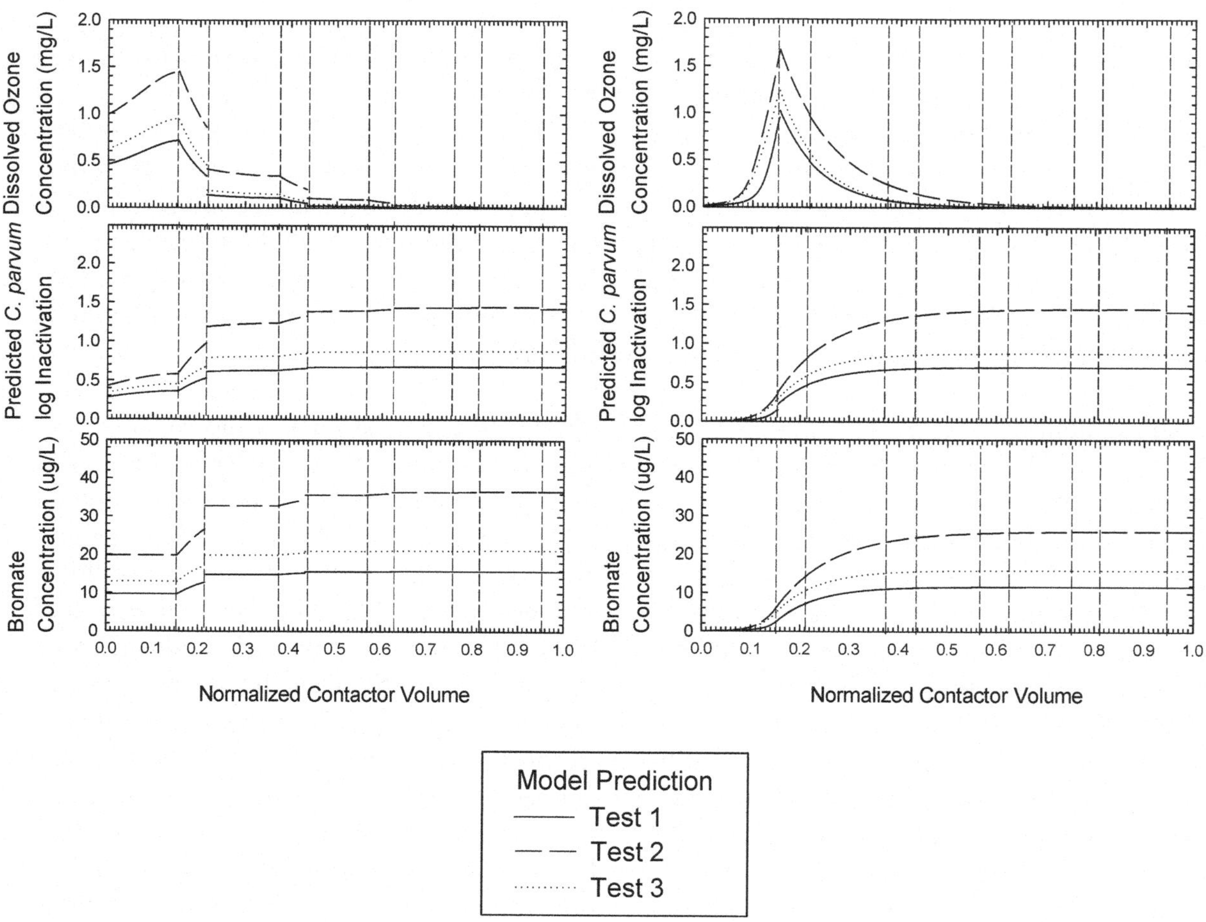


**Figure 4.18: Comparing model simulations of dissolved ozone concentration, *C. parvum* inactivation, and bromate concentration profiles for tests 1 to 3 of Phase II experiments for actual (left) and PFR (right) hydrodynamic conditions**

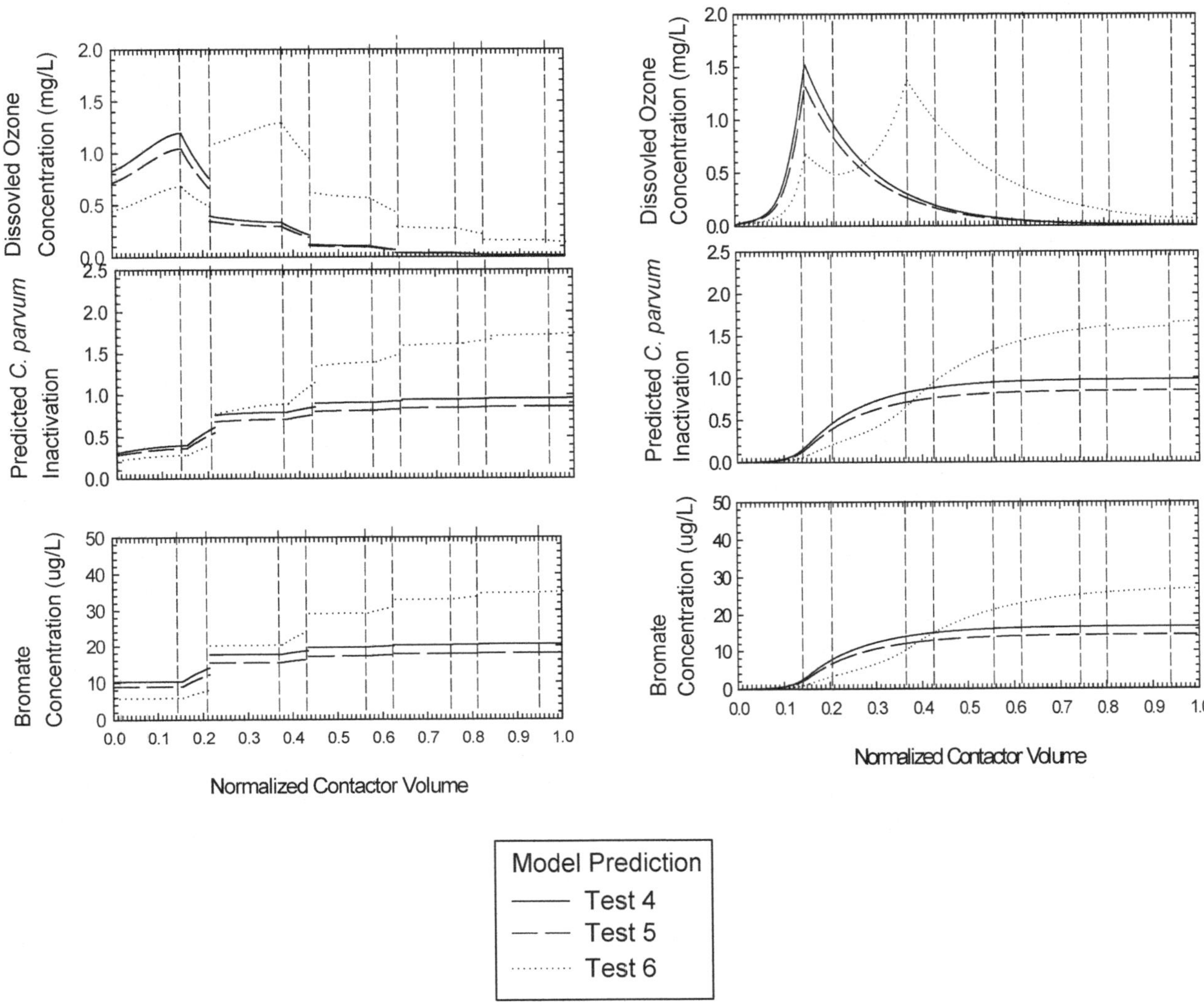


**Figure 4.19: Comparing model simulations of dissolved ozone concentration, *C. parvum* inactivation, and bromate concentration profiles for tests 4 to 6 of Phase II experiments for actual (left) and PFR (right) hydrodynamic conditions**

# CHAPTER 5. SUMMARY AND CONCLUSION

The design and operation of ozone disinfection systems needs to be optimized in order to provide adequate inactivation of *C. parvum* inactivation, while minimizing the formation of disinfection by-products (DBPs) such as bromate. However, while measuring bromate formation is a relatively easy task, assessment of *C. parvum* disinfection efficiency is practically impossible in full-scale ozone contactors, due to the low density of pathogens in natural waters. The *CT* approach, as specified by the Surface Water Treatment Rule (SWTR), may have limitations and uncertainties, due to the complex hydrodynamics and non-idealities in gas and water flow within an ozone contactor. The use of fluorescent-dyed microspheres as surrogates for *C. parvum* oocysts is proposed as an alternative approach. This concept consists of exposing fluorescent-dyed microspheres to the same hydrodynamic environment experienced by the pathogens. The fluorescence intensity of the microspheres would decrease as a result of the exposure to ozone. This loss of fluorescence can then be correlated to the inactivation kinetics of specific pathogens.

Efforts were made to develop a microsphere with Rhodamine WT, a dye currently certified for use in drinking water treatment facilities. One major challenge associated to the use of Rhodamine WT is its relatively high solubility in water, and corresponding tendency to leach out from the polystyrene matrix of the particle. This problem was solved by binding the carboxylic group of the dye molecule to the amine-group on the surface of modified polystyrene microspheres. Unfortunately, the resulting fluorescence intensity of this first generation of Rhodamine WT microspheres was similar to that of the background fluorescence in the raw water tested. Such impasse could be avoided by increasing the mass of Rhodamine WT by allowing the dye to penetrate deeper into the microsphere prior to cross-linking. The limited resources available for this project did not allow the undertaking of such second phase of Rhodamine WT microsphere development.

Of two types of microspheres tested at the full-scale, Fluoresbrite™ Plain YG microspheres (Polysciences, Inc., Warrington, PA) provided the best surrogate. The microspheres were first optimized by preozonation, which reduced the fluorescence intensity to about 5% of the original value. Then, fluorescence decay kinetics of preozonated microspheres were obtained by performing semi-batch experiments. The fluorescence intensity of the microspheres was quantified by flow cytometry, a technique capable of measuring the fluorescence intensity of individual microsphere. The correlation between fluorescence decay and *C. parvum* inactivation was developed by the threshold concept (Chiou *et al.*, 1997), where microspheres above and below a specific fluorescence intensity threshold are considered viable and non-viable oocysts, respectively. Such correlations were used to predict the level of *C. parvum* inactivation achieved in full-scale experiments. Batch ozonation experiments were also performed to determine the parameters for ozone decomposition kinetics for both Phase I and Phase II ACWD raw waters.

Mathematical model predictions of ozone concentration and *C. parvum* inactivation profiles through the full-scale contactor were significantly higher than the Phase I experimental data. The suspected causes for such discrepancy were changes in water quality parameters during sample storage, and clumping of microspheres.

Batch ozonation experiments revealed the occurrence of a linear dependence of bromate formation on *CT* parameter. The mathematical model prediction of ozone concentration, *C. parvum* inactivation, and bromate formation matched the experimental data reasonably well.

Mathematical model simulations reveal that the formation of bromate could be lowered by contactor design and configuration changes that would result in a decrease in the level of back-mixing taking place in contactor chambers. Potential contactor design modifications aimed at minimizing dispersion include changing the configuration of diffuser, and adding vertical baffles.

# CHAPTER 6. RECOMMENDATIONS TO THE INDUSTRY

This study has shown that microspheres can be used as a surrogate for assessing the inactivation efficiency of an ozone contactor. It is therefore recommended that water utilities using ozone as disinfectant consider performing such test. Utilities need to check with their regulatory agencies to determine if using the microspheres for such an application would be acceptable before proceeding with the study.

This study has also shown that mathematical modeling could be used to predict the performance the contactor, if the ozone decay parameter in the specific raw water is available. Mathematical simulation revealed that the hydrodynamics of the contactor has a significant effect on *C. parvum* inactivation, but has virtually no effect on bromate formation. A PFR-like contactor would provide more inactivation for *C. parvum* inactivation than the CSTR-like contactor, while the two types of behavior would cause the same bromate formation. It is therefore recommended to modify the contactor to minimize forward dispersion or in the direction of flow which can be achieved in different ways: by modifying the configuration of the diffuser, by using a diffuser rod that would produce smaller gas bubbles, or by adding vertical baffles in the contactor.

# APPENDIX
# MICROSPHERE DATA

**Phase I Semi-Batch Experiment**

**Pre-ozonated 0.93 μm YG-Microspheres**

**Temp.** 22.3 °C
**pH** 7
**[$O_3$]** 0.978 mg/L

| | *CT* (mg-min/L) | | | | | |
|---|---|---|---|---|---|---|
| **Intensity** | **0** | **1** | **2** | **3** | **4** | **5** |
| **1** | 0 | 0 | 0 | 0 | 0 | 0 |
| **2** | 0 | 0 | 0 | 0 | 0 | 0 |
| **3** | 0 | 0 | 0 | 0 | 0 | 0 |
| **4** | 0 | 0 | 0 | 0 | 0 | 0 |
| **5** | 0 | 0 | 0 | 0 | 0 | 0 |
| **6** | 0 | 0 | 0 | 0 | 0 | 0 |
| **7** | 0 | 0 | 0 | 0 | 0 | 0 |
| **8** | 0 | 0 | 0 | 0 | 0 | 0 |
| **9** | 0 | 0 | 0 | 0 | 0 | 0 |
| **10** | 0 | 0 | 0 | 0 | 0 | 0 |
| **11** | 0 | 0 | 3 | 0 | 3 | 10 |
| **12** | 0 | 3 | 3 | 15 | 37 | 43 |
| **13** | 1 | 5 | 13 | 44 | 50 | 115 |
| **14** | 3 | 8 | 32 | 54 | 103 | 301 |
| **15** | 3 | 28 | 44 | 94 | 166 | 585 |
| **16** | 1 | 30 | 67 | 122 | 311 | 900 |
| **17** | 2 | 40 | 60 | 156 | 542 | 1274 |
| **18** | 1 | 46 | 95 | 314 | 920 | 1554 |
| **19** | 1 | 52 | 139 | 566 | 1278 | 1502 |
| **20** | 1 | 80 | 232 | 837 | 1483 | 1320 |
| **21** | 1 | 75 | 413 | 1177 | 1437 | 890 |
| **22** | 4 | 96 | 623 | 1359 | 1209 | 595 |
| **23** | 5 | 162 | 950 | 1415 | 884 | 354 |
| **24** | 2 | 194 | 1164 | 1165 | 576 | 201 |
| **25** | 6 | 310 | 1267 | 848 | 336 | 102 |
| **26** | 6 | 480 | 1143 | 615 | 226 | 50 |
| **27** | 9 | 648 | 1015 | 379 | 115 | 34 |
| **28** | 17 | 793 | 772 | 232 | 79 | 17 |
| **29** | 20 | 992 | 569 | 179 | 51 | 14 |
| **30** | 24 | 1017 | 402 | 113 | 25 | 6 |
| **31** | 27 | 1066 | 263 | 58 | 9 | 5 |
| **32** | 25 | 926 | 194 | 22 | 8 | 1 |
| **33** | 32 | 783 | 135 | 19 | 3 | 1 |
| **34** | 29 | 596 | 68 | 10 | 4 | 1 |
| **35** | 37 | 437 | 54 | 8 | 1 | 0 |
| **36** | 43 | 280 | 20 | 10 | 1 | 0 |
| **37** | 41 | 198 | 17 | 1 | 0 | 0 |
| **38** | 44 | 108 | 13 | 2 | 0 | 0 |
| **39** | 48 | 75 | 2 | 2 | 0 | 0 |
| **40** | 59 | 55 | 4 | 0 | 1 | 0 |

| | *CT* (mg-min/L) | | | | | |
|---|---|---|---|---|---|---|
| **Intensity** | **0** | **1** | **2** | **3** | **4** | **5** |
| **41** | 51 | 27 | 3 | 0 | 0 | 0 |
| **42** | 70 | 21 | 3 | 0 | 0 | 0 |
| **43** | 96 | 20 | 1 | 0 | 0 | 0 |
| **44** | 121 | 10 | 1 | 0 | 0 | 0 |
| **45** | 197 | 7 | 0 | 0 | 0 | 0 |
| **46** | 279 | 4 | 0 | 0 | 0 | 0 |
| **47** | 317 | 3 | 0 | 0 | 0 | 0 |
| **48** | 432 | 4 | 0 | 1 | 0 | 0 |
| **49** | 536 | 1 | 0 | 0 | 0 | 0 |
| **50** | 638 | 2 | 0 | 0 | 0 | 0 |
| **51** | 766 | 4 | 0 | 0 | 0 | 0 |
| **52** | 786 | 1 | 0 | 0 | 0 | 0 |
| **53** | 796 | 4 | 0 | 0 | 0 | 0 |
| **54** | 739 | 0 | 0 | 0 | 0 | 0 |
| **55** | 682 | 0 | 0 | 1 | 0 | 0 |
| **56** | 575 | 1 | 0 | 0 | 0 | 0 |
| **57** | 514 | 0 | 0 | 0 | 1 | 0 |
| **58** | 370 | 1 | 0 | 0 | 0 | 0 |
| **59** | 277 | 0 | 0 | 0 | 0 | 0 |
| **60** | 245 | 0 | 0 | 0 | 0 | 0 |
| **61** | 165 | 0 | 0 | 0 | 0 | 0 |
| **62** | 118 | 1 | 0 | 0 | 0 | 0 |
| **63** | 80 | 0 | 0 | 0 | 0 | 0 |
| **64** | 60 | 0 | 0 | 0 | 0 | 0 |
| **65** | 45 | 0 | 0 | 0 | 0 | 0 |
| **66** | 29 | 0 | 0 | 0 | 0 | 0 |
| **67** | 34 | 0 | 0 | 0 | 0 | 0 |
| **68** | 21 | 0 | 0 | 0 | 0 | 0 |
| **69** | 18 | 0 | 0 | 0 | 0 | 0 |
| **70** | 15 | 0 | 0 | 0 | 0 | 0 |
| **71** | 12 | 0 | 0 | 0 | 0 | 0 |
| **72** | 8 | 0 | 0 | 0 | 0 | 0 |
| **73** | 4 | 0 | 0 | 0 | 0 | 0 |
| **74** | 5 | 0 | 0 | 0 | 0 | 0 |
| **75** | 4 | 0 | 0 | 0 | 0 | 0 |
| **76** | 5 | 0 | 0 | 0 | 0 | 0 |
| **77** | 1 | 0 | 0 | 0 | 0 | 0 |
| **78** | 3 | 0 | 0 | 0 | 0 | 0 |
| **79** | 0 | 0 | 0 | 0 | 0 | 0 |
| **80** | | | | | | |

**Phase I Batch Experiments**

**Temp.** 20 °C

**Test 1**

| time (min) | $[O_3]$ (mg/L) |
|---|---|
| 0 | 2.327 |
| 1.33 | 1.269 |
| 3.33 | 0.882 |
| 5.33 | 0.668 |
| 7.33 | 0.523 |
| 9.33 | 0.423 |
| 11.33 | 0.354 |

**Test 2**

| time (min) | $[O_3]$ (mg/L) |
|---|---|
| 0 | 1.937 |
| 1.33 | 0.864 |
| 3.33 | 0.523 |
| 5.33 | 0.343 |
| 7.33 | 0.257 |
| 9.33 | 0.176 |
| 11.33 | 0.120 |

**Test 3**

| time (min) | $[O_3]$ (mg/L) |
|---|---|
| 0 | 0.985 |
| 1.33 | 0.254 |
| 3.33 | 0.096 |
| 5.33 | 0.081 |
| 7.33 | 0.076 |

Phase I Full-Scale Experiment

Dissolved Ozone Concentration (mg/L)

| Chamber # | Port # | Test 1 | Test 2 | Test 3 | Test 4 | Test 5 | Test 6 |
|---|---|---|---|---|---|---|---|
| 1 | 1 | 0.060475 | 0.190065 | 0.051836 | 0.276458 | 0.060475 | 0.190065 |
| | 2 | 0.138229 | 0.172786 | 0 | 0.207343 | 0.017279 | 0.12095 |
| | 3 | 0.095032 | 0.207343 | 0.077754 | 0.259179 | 0 | 0.138229 |
| | 4 | 0.103672 | 0.362851 | 0.051836 | 0.285097 | 0.034557 | 0.181425 |
| | 5 | 0.086393 | 0.319654 | 0.112311 | 0.328294 | 0.017279 | 0.086393 |
| | b | 0.112311 | 0.475162 | 0.043197 | 0.587473 | 0.043197 | 0.112311 |
| 2 | 1 | 0.207343 | 0.907127 | 0.086393 | 0.509719 | 0.034557 | 0.259179 |
| | 2 | 0.12959 | 0.933045 | 0.103672 | 0.62203 | 0.060475 | 0.345572 |
| | 3 | 0.397408 | 1.088553 | 0.336933 | 0.976242 | 0.164147 | 0.639309 |
| | 4 | 0.604752 | 0.924406 | 0.509719 | 1.105832 | 0.207343 | 0.665227 |
| | 5 | 0.526998 | 0.863931 | 0.354212 | 0.87257 | 0.267819 | 0.552916 |
| | b | | | | | | |
| 3 | 1 | 0.181425 | 0.466523 | 0.12095 | 0.526998 | 0 | 0.224622 |
| | 2 | 0.215983 | 0.518359 | 0.095032 | 0.492441 | 0 | 0.25054 |
| | 3 | 0.215983 | 0.457883 | 0.12095 | 0.535637 | 0.017279 | 0.181425 |
| | 4 | 0.233261 | 0.475162 | 0.077754 | 0.526998 | 0.017279 | 0.215983 |
| | 5 | 0.190065 | 0.50108 | 0.112311 | 0.466523 | 0.017279 | 0.198704 |
| | b | 0.155508 | 0.311015 | 0.017279 | 0.406048 | 0.017279 | 0.12959 |
| 4 | 1 | 0.095032 | 0.25054 | 0 | 0.259179 | 0 | 0.060475 |
| | 2 | 0.051836 | 0.267819 | 0 | 0.190065 | 0 | 0.017279 |
| | 3 | 0.069114 | 0.285097 | 0.017279 | 0.293737 | 0 | 0.077754 |
| | 4 | 0.069114 | 0.293737 | 0 | 0.224622 | 0 | 0.034557 |
| | 5 | 0.086393 | 0.259179 | 0 | 0.259179 | 0 | 0.069114 |
| | b | 0.017279 | 0.25054 | 0.017279 | 0.207343 | 0 | 0 |
| 5 | 1 | 0 | 0.155508 | 0 | 0.172786 | 0 | 0 |
| | 2 | 0 | 0.12959 | 0 | 0.069114 | 0 | 0 |
| | 3 | 0 | 0.12095 | 0 | 0.172786 | 0 | 0 |
| | 4 | 0 | 0.146868 | 0 | 0.164147 | 0 | 0 |
| | 5 | 0 | 0.12095 | 0 | 0.138229 | 0 | 0 |
| | b | 0 | 0.146868 | 0 | 0.12959 | 0 | 0 |

**Phase I Full-Scale Experiment**
**Test #** 1 **Chamber #** 3
**Pre-ozonated 0.93 μm YG-Microspheres**

| Intensity | Port #1 | Port #2 | Port #3 | Port #4 | Port #5 | Port #b |
|---|---|---|---|---|---|---|
| 1 | | | 382 | 0 | | 0 |
| 2 | | 0 | 813 | 0 | | 0 |
| 3 | | 0 | 652 | 0 | | 1 |
| 4 | | 15 | 352 | 6 | | 4 |
| 5 | | 135 | 155 | 64 | | 36 |
| 6 | | 300 | 68 | 127 | | 61 |
| 7 | | 267 | 29 | 133 | | 54 |
| 8 | | 126 | 12 | 53 | | 33 |
| 9 | | 47 | 7 | 29 | | 13 |
| 10 | | 22 | 7 | 12 | | 11 |
| 11 | | 12 | 4 | 3 | | 18 |
| 12 | | 20 | 3 | 5 | | 24 |
| 13 | | 38 | 3 | 8 | | 37 |
| 14 | | 35 | 3 | 20 | | 60 |
| 15 | | 39 | 5 | 21 | | 77 |
| 16 | | 36 | 7 | 14 | | 87 |
| 17 | | 41 | 1 | 26 | | 118 |
| 18 | | 26 | 10 | 22 | | 128 |
| 19 | | 26 | 7 | 36 | | 175 |
| 20 | | 35 | 6 | 50 | | 205 |
| 21 | | 32 | 13 | 74 | | 230 |
| 22 | | 38 | 11 | 74 | | 249 |
| 23 | | 53 | 21 | 82 | | 256 |
| 24 | | 50 | 16 | 108 | | 227 |
| 25 | | 47 | 22 | 123 | | 240 |
| 26 | | 93 | 30 | 150 | | 184 |
| 27 | | 68 | 46 | 170 | | 182 |
| 28 | | 98 | 43 | 155 | | 101 |
| 29 | | 105 | 43 | 142 | | 78 |
| 30 | | 123 | 46 | 142 | | 45 |
| 31 | | 125 | 54 | 146 | | 40 |
| 32 | | 145 | 34 | 126 | | 34 |
| 33 | | 175 | 48 | 115 | | 19 |
| 34 | | 165 | 33 | 98 | | 15 |
| 35 | | 178 | 41 | 80 | | 4 |
| 36 | | 167 | 31 | 62 | | 5 |
| 37 | | 163 | 33 | 31 | | 3 |
| 38 | | 147 | 39 | 28 | | 4 |
| 39 | | 138 | 16 | 20 | | 2 |
| 40 | | 105 | 15 | 17 | | 2 |

| Intensity | Port #1 | Port #2 | Port #3 | Port #4 | Port #5 | Port #b |
|---|---|---|---|---|---|---|
| 41 | | 98 | 14 | 19 | | 0 |
| 42 | | 80 | 10 | 5 | | 2 |
| 43 | | 84 | 12 | 6 | | 0 |
| 44 | | 55 | 3 | 2 | | 0 |
| 45 | | 40 | 2 | 5 | | 2 |
| 46 | | 30 | 7 | 1 | | 1 |
| 47 | | 17 | 2 | 1 | | 1 |
| 48 | | 16 | 1 | 4 | | 0 |
| 49 | | 16 | 4 | 2 | | 1 |
| 50 | | 10 | 3 | 1 | | 2 |
| 51 | | 5 | 2 | 1 | | 1 |
| 52 | | 3 | 0 | 1 | | 1 |
| 53 | | 5 | 2 | 0 | | 2 |
| 54 | | 2 | 1 | 1 | | 0 |
| 55 | | 0 | 2 | 1 | | 1 |
| 56 | | 5 | 0 | 0 | | 0 |
| 57 | | 3 | 0 | 2 | | 1 |
| 58 | | 3 | 0 | 2 | | 0 |
| 59 | | 2 | 0 | 0 | | 2 |
| 60 | | 1 | 0 | 0 | | 0 |
| 61 | | 2 | 0 | 2 | | 1 |
| 62 | | 1 | 0 | 1 | | 0 |
| 63 | | 0 | 0 | 0 | | 0 |
| 64 | | 1 | 0 | 0 | | 0 |
| 65 | | 1 | 0 | 0 | | 0 |
| 66 | | 1 | 0 | 0 | | 0 |
| 67 | | 0 | 0 | 0 | | 0 |
| 68 | | 0 | 0 | 0 | | 0 |
| 69 | | 0 | 0 | 0 | | 0 |
| 70 | | 2 | 0 | 1 | | 2 |
| 71 | | 0 | 0 | 2 | | 0 |
| 72 | | 1 | 0 | 0 | | 0 |
| 73 | | 1 | 0 | 0 | | 0 |
| 74 | | 0 | 0 | 0 | | 0 |
| 75 | | 1 | 0 | 0 | | 0 |
| 76 | | 2 | 0 | 0 | | 0 |
| 77 | | 0 | 0 | 0 | | 0 |
| 78 | | 0 | 0 | 0 | | 1 |
| 79 | | 0 | 0 | 0 | | 0 |
| 80 | | 0 | 0 | 0 | | 0 |

**Phase I Full-Scale Experiment**
**Test #** 1 **Chamber #** 4
**Pre-ozonated 0.93 μm YG-Microspheres**

| Intensity | Port #1 | Port #2 | Port #3 | Port #4 | Port #5 | Port #b |
|---|---|---|---|---|---|---|
| **1** | 0 | 0 | 1524 | 4100 | 5954 | |
| **2** | 0 | 0 | 3611 | 8578 | 13455 | |
| **3** | 0 | 0 | 2625 | 5820 | 10649 | |
| **4** | 1 | 0 | 1293 | 2380 | 5164 | |
| **5** | 31 | 4 | 581 | 917 | 2388 | |
| **6** | 82 | 7 | 282 | 420 | 1097 | |
| **7** | 59 | 10 | 160 | 190 | 500 | |
| **8** | 23 | 5 | 79 | 95 | 314 | |
| **9** | 6 | 1 | 63 | 55 | 195 | |
| **10** | 0 | 1 | 51 | 30 | 126 | |
| **11** | 5 | 2 | 42 | 17 | 109 | |
| **12** | 5 | 5 | 38 | 22 | 75 | |
| **13** | 15 | 10 | 38 | 16 | 69 | |
| **14** | 25 | 12 | 38 | 11 | 61 | |
| **15** | 28 | 16 | 32 | 15 | 55 | |
| **16** | 39 | 9 | 30 | 18 | 41 | |
| **17** | 47 | 13 | 35 | 20 | 51 | |
| **18** | 47 | 17 | 51 | 20 | 33 | |
| **19** | 68 | 28 | 62 | 33 | 48 | |
| **20** | 77 | 21 | 55 | 34 | 52 | |
| **21** | 105 | 23 | 64 | 41 | 49 | |
| **22** | 112 | 31 | 90 | 38 | 60 | |
| **23** | 159 | 42 | 92 | 79 | 61 | |
| **24** | 183 | 53 | 155 | 85 | 69 | |
| **25** | 215 | 63 | 176 | 89 | 79 | |
| **26** | 237 | 71 | 222 | 106 | 108 | |
| **27** | 235 | 52 | 250 | 134 | 99 | |
| **28** | 227 | 70 | 279 | 154 | 144 | |
| **29** | 259 | 74 | 288 | 181 | 119 | |
| **30** | 295 | 95 | 338 | 186 | 159 | |
| **31** | 272 | 88 | 364 | 195 | 157 | |
| **32** | 274 | 82 | 334 | 198 | 137 | |
| **33** | 253 | 71 | 327 | 232 | 165 | |
| **34** | 231 | 56 | 290 | 219 | 157 | |
| **35** | 156 | 69 | 251 | 233 | 130 | |
| **36** | 162 | 43 | 209 | 199 | 137 | |
| **37** | 108 | 32 | 211 | 196 | 122 | |
| **38** | 100 | 24 | 117 | 165 | 100 | |
| **39** | 105 | 15 | 104 | 165 | 84 | |
| **40** | 98 | 15 | 71 | 134 | 70 | |

| Intensity | Port #1 | Port #2 | Port #3 | Port #4 | Port #5 | Port #b |
|---|---|---|---|---|---|---|
| **41** | 76 | 16 | 71 | 112 | 56 | |
| **42** | 94 | 4 | 45 | 98 | 34 | |
| **43** | 95 | 6 | 24 | 76 | 39 | |
| **44** | 74 | 7 | 20 | 52 | 35 | |
| **45** | 85 | 3 | 17 | 44 | 16 | |
| **46** | 89 | 0 | 17 | 32 | 18 | |
| **47** | 83 | 3 | 10 | 29 | 10 | |
| **48** | 74 | 1 | 8 | 18 | 10 | |
| **49** | 88 | 1 | 11 | 19 | 5 | |
| **50** | 73 | 3 | 12 | 9 | 5 | |
| **51** | 64 | 3 | 9 | 17 | 3 | |
| **52** | 62 | 0 | 14 | 7 | 6 | |
| **53** | 43 | 2 | 5 | 3 | 4 | |
| **54** | 28 | 0 | 10 | 1 | 3 | |
| **55** | 24 | 0 | 10 | 2 | 4 | |
| **56** | 19 | 1 | 6 | 3 | 1 | |
| **57** | 15 | 3 | 7 | 2 | 4 | |
| **58** | 9 | 0 | 5 | 1 | 1 | |
| **59** | 12 | 0 | 5 | 0 | 4 | |
| **60** | 3 | 0 | 4 | 1 | 2 | |
| **61** | 4 | 0 | 2 | 3 | 1 | |
| **62** | 7 | 0 | 2 | 0 | 0 | |
| **63** | 1 | 0 | 3 | 0 | 2 | |
| **64** | 0 | 0 | 1 | 0 | 2 | |
| **65** | 2 | 0 | 1 | 0 | 5 | |
| **66** | 0 | 0 | 1 | 1 | 2 | |
| **67** | 0 | 1 | 2 | 0 | 3 | |
| **68** | 0 | 0 | 0 | 0 | 1 | |
| **69** | 0 | 0 | 0 | 0 | 1 | |
| **70** | 0 | 0 | 1 | 0 | 2 | |
| **71** | 0 | 0 | 0 | 0 | 0 | |
| **72** | 0 | 0 | 1 | 1 | 2 | |
| **73** | 0 | 0 | 1 | 0 | 4 | |
| **74** | 0 | 0 | 2 | 0 | 0 | |
| **75** | 0 | 0 | 0 | 0 | 0 | |
| **76** | 0 | 0 | 1 | 0 | 1 | |
| **77** | 0 | 0 | 0 | 0 | 1 | |
| **78** | 0 | 0 | 2 | 0 | 1 | |
| **79** | 0 | 0 | 0 | 0 | 0 | |
| **80** | 0 | 0 | 0 | 0 | 0 | |

**Phase I Full-Scale Experiment**
**Test #** 2 **Chamber #** 1
**Pre-ozonated 0.93 μm YG-Microspheres**

| Intensity | Port #1 | Port #2 | Port #3 | Port #4 | Port #5 | Port #b |
|---|---|---|---|---|---|---|
| 1 | 1973 | 1038 | 214 | 224 | 178 | 97 |
| 2 | 4665 | 2563 | 524 | 473 | 429 | 247 |
| 3 | 5165 | 2486 | 477 | 409 | 438 | 342 |
| 4 | 2966 | 1188 | 263 | 210 | 206 | 326 |
| 5 | 1178 | 411 | 101 | 109 | 103 | 203 |
| 6 | 424 | 95 | 52 | 63 | 60 | 151 |
| 7 | 195 | 36 | 36 | 42 | 32 | 90 |
| 8 | 114 | 19 | 22 | 25 | 23 | 46 |
| 9 | 60 | 8 | 19 | 19 | 21 | 43 |
| 10 | 55 | 5 | 11 | 18 | 26 | 24 |
| 11 | 28 | 2 | 10 | 11 | 15 | 27 |
| 12 | 23 | 2 | 11 | 7 | 14 | 23 |
| 13 | 19 | 4 | 14 | 7 | 9 | 18 |
| 14 | 14 | 5 | 5 | 9 | 11 | 22 |
| 15 | 14 | 4 | 9 | 10 | 21 | 17 |
| 16 | 17 | 7 | 7 | 11 | 11 | 27 |
| 17 | 18 | 7 | 12 | 12 | 12 | 23 |
| 18 | 22 | 8 | 8 | 17 | 13 | 36 |
| 19 | 31 | 7 | 22 | 16 | 23 | 33 |
| 20 | 34 | 8 | 19 | 18 | 25 | 27 |
| 21 | 35 | 12 | 20 | 22 | 18 | 43 |
| 22 | 38 | 10 | 19 | 25 | 14 | 57 |
| 23 | 49 | 19 | 25 | 35 | 19 | 58 |
| 24 | 46 | 26 | 26 | 26 | 41 | 63 |
| 25 | 61 | 29 | 38 | 31 | 34 | 75 |
| 26 | 73 | 35 | 48 | 47 | 39 | 94 |
| 27 | 61 | 30 | 52 | 48 | 67 | 79 |
| 28 | 79 | 33 | 56 | 58 | 90 | 85 |
| 29 | 95 | 52 | 69 | 59 | 63 | 80 |
| 30 | 77 | 54 | 63 | 70 | 77 | 90 |
| 31 | 84 | 52 | 81 | 80 | 94 | 96 |
| 32 | 105 | 60 | 81 | 66 | 83 | 93 |
| 33 | 106 | 70 | 76 | 88 | 93 | 106 |
| 34 | 95 | 76 | 81 | 96 | 110 | 121 |
| 35 | 125 | 75 | 106 | 80 | 97 | 106 |
| 36 | 93 | 93 | 96 | 110 | 97 | 119 |
| 37 | 94 | 90 | 118 | 86 | 105 | 104 |
| 38 | 73 | 99 | 92 | 118 | 131 | 119 |
| 39 | 83 | 85 | 94 | 103 | 128 | 109 |
| 40 | 75 | 86 | 112 | 96 | 119 | 101 |

| Intensity | Port #1 | Port #2 | Port #3 | Port #4 | Port #5 | Port #b |
|---|---|---|---|---|---|---|
| 41 | 66 | 103 | 122 | 110 | 133 | 99 |
| 42 | 62 | 89 | 122 | 94 | 109 | 90 |
| 43 | 58 | 79 | 116 | 90 | 107 | 90 |
| 44 | 53 | 88 | 86 | 86 | 103 | 74 |
| 45 | 46 | 78 | 92 | 82 | 95 | 87 |
| 46 | 37 | 75 | 93 | 94 | 64 | 63 |
| 47 | 32 | 67 | 81 | 74 | 77 | 58 |
| 48 | 22 | 79 | 76 | 80 | 90 | 60 |
| 49 | 18 | 68 | 67 | 84 | 71 | 46 |
| 50 | 15 | 58 | 64 | 77 | 60 | 37 |
| 51 | 16 | 62 | 62 | 53 | 70 | 27 |
| 52 | 13 | 61 | 56 | 62 | 59 | 26 |
| 53 | 4 | 41 | 46 | 53 | 29 | 19 |
| 54 | 6 | 38 | 46 | 46 | 41 | 20 |
| 55 | 7 | 29 | 28 | 41 | 27 | 11 |
| 56 | 3 | 26 | 21 | 42 | 19 | 19 |
| 57 | 0 | 22 | 35 | 31 | 14 | 9 |
| 58 | 2 | 14 | 29 | 16 | 8 | 9 |
| 59 | 0 | 14 | 26 | 12 | 11 | 4 |
| 60 | 1 | 8 | 20 | 13 | 12 | 6 |
| 61 | 0 | 10 | 13 | 10 | 2 | 2 |
| 62 | 2 | 8 | 7 | 6 | 2 | 2 |
| 63 | 1 | 6 | 5 | 6 | 1 | 2 |
| 64 | 0 | 4 | 9 | 5 | 0 | 1 |
| 65 | 0 | 3 | 5 | 7 | 0 | 2 |
| 66 | 0 | 1 | 1 | 2 | 1 | 1 |
| 67 | 2 | 2 | 2 | 1 | 0 | 1 |
| 68 | 0 | 0 | 0 | 0 | 1 | 1 |
| 69 | 2 | 2 | 2 | 0 | 0 | 0 |
| 70 | 2 | 2 | 1 | 0 | 1 | 1 |
| 71 | 1 | 0 | 1 | 0 | 1 | 1 |
| 72 | 0 | 1 | 0 | 1 | 0 | 0 |
| 73 | 0 | 0 | 0 | 0 | 0 | 0 |
| 74 | 1 | 0 | 1 | 0 | 0 | 0 |
| 75 | 0 | 0 | 0 | 0 | 0 | 0 |
| 76 | 0 | 0 | 0 | 0 | 0 | 1 |
| 77 | 0 | 0 | 0 | 0 | 0 | 1 |
| 78 | 0 | 0 | 0 | 0 | 0 | 1 |
| 79 | 1 | 0 | 0 | 0 | 0 | 0 |
| 80 | 0 | 0 | 0 | 0 | 0 | 0 |

**Phase I Full-Scale Experiment**

**Test #** 2 **Chamber #** 2

**Pre-ozonated 0.93 μm YG-Microspheres**

| Intensity | Port #1 | Port #2 | Port #3 | Port #4 | Port #5 | Port #b |
|---|---|---|---|---|---|---|
| **1** | 134 | 97 | 56 | 436 | 89 | |
| **2** | 329 | 263 | 176 | 980 | 199 | |
| **3** | 292 | 322 | 222 | 1024 | 228 | |
| **4** | 154 | 235 | 154 | 677 | 189 | |
| **5** | 72 | 153 | 81 | 385 | 100 | |
| **6** | 40 | 140 | 42 | 203 | 41 | |
| **7** | 22 | 85 | 19 | 146 | 36 | |
| **8** | 14 | 79 | 15 | 93 | 25 | |
| **9** | 28 | 58 | 9 | 52 | 16 | |
| **10** | 25 | 45 | 16 | 52 | 27 | |
| **11** | 18 | 54 | 17 | 47 | 30 | |
| **12** | 34 | 37 | 28 | 38 | 30 | |
| **13** | 31 | 38 | 31 | 29 | 46 | |
| **14** | 49 | 25 | 38 | 35 | 34 | |
| **15** | 40 | 45 | 32 | 47 | 51 | |
| **16** | 64 | 34 | 34 | 45 | 49 | |
| **17** | 87 | 52 | 57 | 55 | 72 | |
| **18** | 75 | 43 | 55 | 66 | 63 | |
| **19** | 101 | 61 | 74 | 63 | 78 | |
| **20** | 97 | 78 | 86 | 86 | 81 | |
| **21** | 112 | 47 | 89 | 102 | 98 | |
| **22** | 124 | 58 | 109 | 98 | 99 | |
| **23** | 114 | 68 | 113 | 99 | 126 | |
| **24** | 118 | 80 | 119 | 112 | 106 | |
| **25** | 126 | 70 | 136 | 114 | 134 | |
| **26** | 137 | 95 | 124 | 109 | 145 | |
| **27** | 145 | 67 | 155 | 127 | 131 | |
| **28** | 130 | 92 | 118 | 121 | 149 | |
| **29** | 147 | 81 | 132 | 112 | 134 | |
| **30** | 139 | 78 | 138 | 108 | 147 | |
| **31** | 143 | 71 | 138 | 112 | 128 | |
| **32** | 104 | 60 | 123 | 109 | 126 | |
| **33** | 110 | 51 | 121 | 100 | 131 | |
| **34** | 87 | 59 | 114 | 118 | 102 | |
| **35** | 82 | 57 | 89 | 112 | 106 | |
| **36** | 92 | 58 | 96 | 111 | 98 | |
| **37** | 61 | 45 | 88 | 87 | 80 | |
| **38** | 77 | 39 | 76 | 77 | 78 | |
| **39** | 70 | 36 | 94 | 72 | 56 | |
| **40** | 59 | 37 | 63 | 65 | 55 | |

| Intensity | Port #1 | Port #2 | Port #3 | Port #4 | Port #5 | Port #b |
|---|---|---|---|---|---|---|
| **41** | 45 | 23 | 60 | 56 | 53 | |
| **42** | 46 | 24 | 58 | 35 | 45 | |
| **43** | 41 | 17 | 48 | 38 | 34 | |
| **44** | 30 | 19 | 31 | 28 | 33 | |
| **45** | 34 | 11 | 31 | 28 | 15 | |
| **46** | 32 | 12 | 29 | 13 | 18 | |
| **47** | 22 | 12 | 16 | 10 | 14 | |
| **48** | 11 | 7 | 16 | 10 | 8 | |
| **49** | 19 | 6 | 8 | 11 | 4 | |
| **50** | 5 | 5 | 9 | 7 | 3 | |
| **51** | 7 | 2 | 8 | 6 | 5 | |
| **52** | 9 | 2 | 9 | 4 | 1 | |
| **53** | 11 | 2 | 1 | 2 | 1 | |
| **54** | 2 | 3 | 1 | 3 | 0 | |
| **55** | 2 | 3 | 0 | 2 | 0 | |
| **56** | 1 | 1 | 2 | 1 | 0 | |
| **57** | 1 | 0 | 1 | 0 | 0 | |
| **58** | 2 | 1 | 0 | 2 | 0 | |
| **59** | 1 | 0 | 0 | 1 | 0 | |
| **60** | 0 | 0 | 0 | 0 | 0 | |
| **61** | 1 | 0 | 0 | 1 | 0 | |
| **62** | 0 | 0 | 0 | 0 | 0 | |
| **63** | 0 | 0 | 0 | 0 | 0 | |
| **64** | 0 | 0 | 0 | 0 | 0 | |
| **65** | 0 | 0 | 0 | 0 | 0 | |
| **66** | 0 | 0 | 0 | 0 | 0 | |
| **67** | 0 | 0 | 0 | 0 | 0 | |
| **68** | 0 | 0 | 0 | 0 | 0 | |
| **69** | 0 | 0 | 0 | 0 | 0 | |
| **70** | 0 | 0 | 0 | 0 | 0 | |
| **71** | 0 | 0 | 0 | 0 | 0 | |
| **72** | 0 | 0 | 0 | 0 | 0 | |
| **73** | 0 | 0 | 0 | 0 | 0 | |
| **74** | 0 | 0 | 0 | 0 | 0 | |
| **75** | 0 | 0 | 0 | 0 | 0 | |
| **76** | 0 | 0 | 0 | 0 | 0 | |
| **77** | 0 | 0 | 0 | 0 | 0 | |
| **78** | 0 | 0 | 0 | 1 | 0 | |
| **79** | 0 | 0 | 0 | 0 | 0 | |
| **80** | 0 | 0 | 0 | 0 | 0 | |

**Phase I Full-Scale Experiment**
**Test #** 2 **Chamber #** 3
**Pre-ozonated 0.93 μm YG-Microspheres**

| Intensity | Port #1 | Port #2 | Port #3 | Port #4 | Port #5 | Port #b |
|---|---|---|---|---|---|---|
| 1 | 493 | 175 | | | 327 | 157 |
| 2 | 1216 | 381 | | | 649 | 357 |
| 3 | 923 | 333 | | | 618 | 293 |
| 4 | 391 | 141 | | | 335 | 151 |
| 5 | 138 | 48 | | | 204 | 85 |
| 6 | 49 | 31 | | | 105 | 40 |
| 7 | 24 | 13 | | | 69 | 41 |
| 8 | 36 | 15 | | | 45 | 39 |
| 9 | 30 | 14 | | | 56 | 36 |
| 10 | 31 | 16 | | | 50 | 49 |
| 11 | 51 | 25 | | | 73 | 65 |
| 12 | 65 | 27 | | | 77 | 79 |
| 13 | 89 | 32 | | | 89 | 106 |
| 14 | 104 | 23 | | | 102 | 103 |
| 15 | 89 | 24 | | | 117 | 135 |
| 16 | 117 | 41 | | | 144 | 176 |
| 17 | 145 | 54 | | | 155 | 183 |
| 18 | 183 | 52 | | | 159 | 189 |
| 19 | 162 | 51 | | | 160 | 201 |
| 20 | 159 | 70 | | | 160 | 241 |
| 21 | 167 | 48 | | | 182 | 240 |
| 22 | 152 | 55 | | | 155 | 223 |
| 23 | 149 | 55 | | | 146 | 188 |
| 24 | 152 | 46 | | | 133 | 186 |
| 25 | 128 | 56 | | | 110 | 218 |
| 26 | 129 | 45 | | | 126 | 154 |
| 27 | 107 | 40 | | | 101 | 161 |
| 28 | 105 | 35 | | | 82 | 139 |
| 29 | 84 | 25 | | | 67 | 100 |
| 30 | 83 | 17 | | | 51 | 85 |
| 31 | 52 | 13 | | | 43 | 64 |
| 32 | 54 | 10 | | | 15 | 40 |
| 33 | 36 | 9 | | | 26 | 28 |
| 34 | 25 | 3 | | | 11 | 20 |
| 35 | 19 | 5 | | | 5 | 21 |
| 36 | 9 | 4 | | | 5 | 10 |
| 37 | 11 | 3 | | | 4 | 8 |
| 38 | 11 | 1 | | | 4 | 7 |
| 39 | 4 | 0 | | | 0 | 3 |
| 40 | 6 | 2 | | | 2 | 8 |

| Intensity | Port #1 | Port #2 | Port #3 | Port #4 | Port #5 | Port #b |
|---|---|---|---|---|---|---|
| 41 | 2 | 0 | | | 1 | 1 |
| 42 | 0 | 2 | | | 0 | 2 |
| 43 | 0 | 0 | | | 0 | 0 |
| 44 | 0 | 1 | | | 0 | 1 |
| 45 | 0 | 0 | | | 0 | 0 |
| 46 | 0 | 0 | | | 1 | 1 |
| 47 | 1 | 0 | | | 0 | 0 |
| 48 | 1 | 0 | | | 1 | 0 |
| 49 | 0 | 0 | | | 0 | 2 |
| 50 | 0 | 0 | | | 0 | 0 |
| 51 | 0 | 0 | | | 0 | 0 |
| 52 | 0 | 0 | | | 0 | 0 |
| 53 | 0 | 0 | | | 0 | 0 |
| 54 | 0 | 0 | | | 0 | 1 |
| 55 | 0 | 0 | | | 0 | 0 |
| 56 | 0 | 0 | | | 0 | 0 |
| 57 | 0 | 0 | | | 0 | 0 |
| 58 | 0 | 0 | | | 0 | 1 |
| 59 | 0 | 0 | | | 0 | 0 |
| 60 | 0 | 0 | | | 0 | 0 |
| 61 | 0 | 0 | | | 0 | 0 |
| 62 | 0 | 0 | | | 0 | 0 |
| 63 | 0 | 0 | | | 0 | 0 |
| 64 | 0 | 0 | | | 0 | 0 |
| 65 | 0 | 0 | | | 1 | 0 |
| 66 | 0 | 0 | | | 1 | 1 |
| 67 | 0 | 0 | | | 0 | 0 |
| 68 | 0 | 0 | | | 0 | 0 |
| 69 | 0 | 1 | | | 0 | 0 |
| 70 | 0 | 0 | | | 0 | 0 |
| 71 | 0 | 0 | | | 0 | 0 |
| 72 | 0 | 0 | | | 0 | 0 |
| 73 | 0 | 0 | | | 0 | 0 |
| 74 | 0 | 0 | | | 0 | 0 |
| 75 | 0 | 0 | | | 0 | 0 |
| 76 | 0 | 0 | | | 0 | 0 |
| 77 | 0 | 0 | | | 0 | 0 |
| 78 | 0 | 0 | | | 0 | 0 |
| 79 | 0 | 0 | | | 1 | 0 |
| 80 | 0 | 0 | | | 0 | 0 |

**Phase I Full-Scale Experiment**
**Test #** 2 **Chamber #** 4
**Pre-ozonated 0.93 μm YG-Microspheres**

| Intensity | Port #1 | Port #2 | Port #3 | Port #4 | Port #5 | Port #b |
|---|---|---|---|---|---|---|
| **1** | | 436 | 271 | 392 | 191 | 475 |
| **2** | | 843 | 586 | 989 | 440 | 1235 |
| **3** | | 648 | 364 | 630 | 304 | 1313 |
| **4** | | 231 | 140 | 310 | 129 | 980 |
| **5** | | 64 | 61 | 125 | 52 | 551 |
| **6** | | 33 | 39 | 82 | 36 | 329 |
| **7** | | 22 | 30 | 61 | 22 | 190 |
| **8** | | 29 | 41 | 45 | 20 | 132 |
| **9** | | 26 | 49 | 42 | 18 | 101 |
| **10** | | 39 | 64 | 46 | 35 | 100 |
| **11** | | 44 | 90 | 52 | 22 | 98 |
| **12** | | 48 | 101 | 69 | 40 | 105 |
| **13** | | 60 | 126 | 74 | 39 | 123 |
| **14** | | 78 | 162 | 67 | 46 | 139 |
| **15** | | 99 | 180 | 100 | 57 | 150 |
| **16** | | 130 | 178 | 98 | 68 | 166 |
| **17** | | 137 | 196 | 117 | 87 | 176 |
| **18** | | 145 | 209 | 116 | 80 | 191 |
| **19** | | 140 | 206 | 118 | 102 | 175 |
| **20** | | 133 | 252 | 139 | 98 | 210 |
| **21** | | 163 | 182 | 107 | 121 | 173 |
| **22** | | 175 | 178 | 130 | 134 | 179 |
| **23** | | 159 | 185 | 125 | 145 | 158 |
| **24** | | 182 | 171 | 107 | 169 | 147 |
| **25** | | 159 | 142 | 109 | 154 | 146 |
| **26** | | 151 | 101 | 88 | 161 | 101 |
| **27** | | 133 | 83 | 75 | 160 | 96 |
| **28** | | 138 | 70 | 69 | 160 | 70 |
| **29** | | 95 | 56 | 40 | 152 | 61 |
| **30** | | 70 | 42 | 39 | 145 | 49 |
| **31** | | 73 | 33 | 22 | 128 | 15 |
| **32** | | 62 | 21 | 16 | 107 | 23 |
| **33** | | 55 | 13 | 10 | 92 | 19 |
| **34** | | 38 | 4 | 4 | 82 | 6 |
| **35** | | 36 | 4 | 9 | 80 | 7 |
| **36** | | 13 | 3 | 6 | 53 | 8 |
| **37** | | 13 | 1 | 3 | 38 | 3 |
| **38** | | 7 | 0 | 3 | 37 | 5 |
| **39** | | 6 | 0 | 2 | 20 | 2 |
| **40** | | 3 | 1 | 1 | 17 | 4 |

| Intensity | Port #1 | Port #2 | Port #3 | Port #4 | Port #5 | Port #b |
|---|---|---|---|---|---|---|
| **41** | | 2 | 0 | 0 | 6 | 2 |
| **42** | | 2 | 0 | 0 | 8 | 2 |
| **43** | | 1 | 0 | 1 | 6 | 2 |
| **44** | | 0 | 0 | 1 | 6 | 0 |
| **45** | | 2 | 0 | 0 | 2 | 2 |
| **46** | | 0 | 0 | 1 | 1 | 0 |
| **47** | | 1 | 1 | 0 | 1 | 2 |
| **48** | | 0 | 0 | 0 | 0 | 0 |
| **49** | | 0 | 0 | 1 | 0 | 0 |
| **50** | | 0 | 0 | 0 | 0 | 1 |
| **51** | | 0 | 0 | 0 | 0 | 0 |
| **52** | | 0 | 0 | 0 | 0 | 2 |
| **53** | | 0 | 0 | 0 | 0 | 0 |
| **54** | | 0 | 0 | 0 | 0 | 0 |
| **55** | | 0 | 0 | 0 | 0 | 1 |
| **56** | | 0 | 0 | 1 | 0 | 1 |
| **57** | | 0 | 0 | 0 | 0 | 0 |
| **58** | | 0 | 0 | 1 | 0 | 0 |
| **59** | | 0 | 0 | 0 | 1 | 0 |
| **60** | | 0 | 0 | 0 | 0 | 0 |
| **61** | | 0 | 0 | 0 | 0 | 0 |
| **62** | | 0 | 0 | 0 | 0 | 0 |
| **63** | | 0 | 0 | 0 | 0 | 1 |
| **64** | | 0 | 0 | 0 | 0 | 0 |
| **65** | | 0 | 0 | 1 | 0 | 3 |
| **66** | | 0 | 0 | 0 | 0 | 1 |
| **67** | | 0 | 0 | 0 | 1 | 0 |
| **68** | | 0 | 0 | 0 | 0 | 0 |
| **69** | | 0 | 0 | 0 | 0 | 0 |
| **70** | | 0 | 0 | 0 | 0 | 0 |
| **71** | | 0 | 0 | 0 | 2 | 0 |
| **72** | | 0 | 0 | 0 | 0 | 0 |
| **73** | | 0 | 0 | 0 | 0 | 0 |
| **74** | | 0 | 0 | 0 | 0 | 1 |
| **75** | | 0 | 0 | 0 | 0 | 0 |
| **76** | | 0 | 0 | 0 | 0 | 0 |
| **77** | | 0 | 0 | 0 | 0 | 2 |
| **78** | | 0 | 0 | 0 | 0 | 0 |
| **79** | | 0 | 0 | 0 | 0 | 0 |
| **80** | | 0 | 0 | 0 | 0 | 0 |

**Phase I Full-Scale Experiment**
**Test #** 2 **Chamber #** 5
**Pre-ozonated 0.93 μm YG-Microspheres**

| Intensity | Port #1 | Port #2 | Port #3 | Port #4 | Port #5 | Port #b |
|---|---|---|---|---|---|---|
| **1** | 176 | 60 | 49 | | 55 | |
| **2** | 428 | 167 | 95 | | 142 | |
| **3** | 370 | 212 | 96 | | 102 | |
| **4** | 185 | 123 | 56 | | 67 | |
| **5** | 90 | 58 | 57 | | 36 | |
| **6** | 58 | 30 | 39 | | 18 | |
| **7** | 56 | 29 | 39 | | 29 | |
| **8** | 50 | 11 | 33 | | 33 | |
| **9** | 72 | 22 | 57 | | 36 | |
| **10** | 83 | 6 | 74 | | 55 | |
| **11** | 99 | 13 | 51 | | 85 | |
| **12** | 153 | 16 | 96 | | 86 | |
| **13** | 162 | 25 | 107 | | 127 | |
| **14** | 184 | 12 | 105 | | 141 | |
| **15** | 196 | 19 | 123 | | 165 | |
| **16** | 206 | 24 | 128 | | 170 | |
| **17** | 222 | 31 | 110 | | 208 | |
| **18** | 237 | 35 | 131 | | 203 | |
| **19** | 226 | 37 | 116 | | 253 | |
| **20** | 188 | 35 | 125 | | 207 | |
| **21** | 167 | 22 | 120 | | 218 | |
| **22** | 181 | 46 | 98 | | 178 | |
| **23** | 147 | 39 | 87 | | 177 | |
| **24** | 117 | 44 | 72 | | 165 | |
| **25** | 88 | 36 | 55 | | 127 | |
| **26** | 80 | 43 | 39 | | 106 | |
| **27** | 69 | 46 | 20 | | 74 | |
| **28** | 39 | 42 | 17 | | 76 | |
| **29** | 29 | 43 | 12 | | 43 | |
| **30** | 9 | 57 | 10 | | 35 | |
| **31** | 12 | 59 | 5 | | 16 | |
| **32** | 13 | 49 | 5 | | 16 | |
| **33** | 7 | 61 | 5 | | 7 | |
| **34** | 3 | 48 | 1 | | 2 | |
| **35** | 4 | 62 | 1 | | 6 | |
| **36** | 1 | 55 | 0 | | 1 | |
| **37** | 0 | 63 | 0 | | 3 | |
| **38** | 0 | 40 | 0 | | 1 | |
| **39** | 2 | 37 | 0 | | 0 | |
| **40** | 1 | 45 | 1 | | 0 | |

| Intensity | Port #1 | Port #2 | Port #3 | Port #4 | Port #5 | Port #b |
|---|---|---|---|---|---|---|
| **41** | 1 | 42 | 0 | | 0 | |
| **42** | 0 | 38 | 0 | | 0 | |
| **43** | 0 | 25 | 1 | | 0 | |
| **44** | 2 | 31 | 0 | | 1 | |
| **45** | 0 | 27 | 1 | | 0 | |
| **46** | 1 | 14 | 0 | | 0 | |
| **47** | 0 | 28 | 0 | | 0 | |
| **48** | 0 | 11 | 0 | | 0 | |
| **49** | 1 | 12 | 0 | | 0 | |
| **50** | 0 | 6 | 0 | | 0 | |
| **51** | 1 | 5 | 0 | | 0 | |
| **52** | 1 | 3 | 0 | | 0 | |
| **53** | 0 | 1 | 1 | | 0 | |
| **54** | 1 | 1 | 0 | | 0 | |
| **55** | 0 | 1 | 0 | | 0 | |
| **56** | 0 | 2 | 0 | | 0 | |
| **57** | 0 | 0 | 0 | | 0 | |
| **58** | 1 | 0 | 0 | | 0 | |
| **59** | 0 | 0 | 0 | | 0 | |
| **60** | 0 | 0 | 0 | | 0 | |
| **61** | 0 | 0 | 0 | | 0 | |
| **62** | 0 | 0 | 0 | | 0 | |
| **63** | 1 | 0 | 0 | | 0 | |
| **64** | 1 | 0 | 0 | | 0 | |
| **65** | 0 | 0 | 0 | | 0 | |
| **66** | 0 | 0 | 0 | | 0 | |
| **67** | 0 | 0 | 0 | | 0 | |
| **68** | 0 | 0 | 0 | | 0 | |
| **69** | 0 | 0 | 0 | | 0 | |
| **70** | 0 | 0 | 0 | | 0 | |
| **71** | 0 | 0 | 0 | | 0 | |
| **72** | 0 | 0 | 0 | | 0 | |
| **73** | 0 | 0 | 0 | | 0 | |
| **74** | 0 | 0 | 0 | | 0 | |
| **75** | 0 | 0 | 1 | | 0 | |
| **76** | 0 | 0 | 1 | | 0 | |
| **77** | 0 | 0 | 0 | | 0 | |
| **78** | 0 | 0 | 0 | | 0 | |
| **79** | 0 | 0 | 0 | | 0 | |
| **80** | 0 | 0 | 0 | | 0 | |

**Phase I Full-Scale Experiment**
**Test #** 3 **Chamber #** 1
**Pre-ozonated 0.93 μm YG-Microspheres**

| Intensity | Port #1 | Port #2 | Port #3 | Port #4 | Port #5 | Port #b |
|---|---|---|---|---|---|---|
| **1** | 67 | 1632 | | | | |
| **2** | 203 | 4418 | | | | |
| **3** | 312 | 5043 | | | | |
| **4** | 245 | 3357 | | | | |
| **5** | 137 | 1591 | | | | |
| **6** | 100 | 809 | | | | |
| **7** | 43 | 436 | | | | |
| **8** | 40 | 318 | | | | |
| **9** | 29 | 195 | | | | |
| **10** | 25 | 125 | | | | |
| **11** | 24 | 91 | | | | |
| **12** | 18 | 78 | | | | |
| **13** | 17 | 62 | | | | |
| **14** | 13 | 36 | | | | |
| **15** | 19 | 35 | | | | |
| **16** | 16 | 14 | | | | |
| **17** | 12 | 13 | | | | |
| **18** | 8 | 9 | | | | |
| **19** | 7 | 10 | | | | |
| **20** | 9 | 8 | | | | |
| **21** | 8 | 8 | | | | |
| **22** | 10 | 13 | | | | |
| **23** | 0 | 13 | | | | |
| **24** | 5 | 21 | | | | |
| **25** | 6 | 15 | | | | |
| **26** | 4 | 22 | | | | |
| **27** | 7 | 30 | | | | |
| **28** | 2 | 37 | | | | |
| **29** | 6 | 42 | | | | |
| **30** | 8 | 51 | | | | |
| **31** | 14 | 70 | | | | |
| **32** | 16 | 50 | | | | |
| **33** | 18 | 77 | | | | |
| **34** | 18 | 100 | | | | |
| **35** | 28 | 94 | | | | |
| **36** | 27 | 88 | | | | |
| **37** | 29 | 106 | | | | |
| **38** | 36 | 108 | | | | |
| **39** | 37 | 103 | | | | |
| **40** | 22 | 96 | | | | |
| **41** | 47 | 138 | | | | |
| **42** | 34 | 124 | | | | |
| **43** | 36 | 118 | | | | |
| **44** | 33 | 122 | | | | |
| **45** | 34 | 99 | | | | |
| **46** | 35 | 102 | | | | |
| **47** | 36 | 94 | | | | |
| **48** | 28 | 94 | | | | |
| **49** | 30 | 78 | | | | |
| **50** | 29 | 71 | | | | |
| **51** | 22 | 76 | | | | |
| **52** | 25 | 59 | | | | |
| **53** | 17 | 56 | | | | |
| **54** | 14 | 60 | | | | |
| **55** | 12 | 52 | | | | |
| **56** | 9 | 40 | | | | |
| **57** | 8 | 30 | | | | |
| **58** | 11 | 32 | | | | |
| **59** | 5 | 13 | | | | |
| **60** | 4 | 14 | | | | |
| **61** | 1 | 19 | | | | |
| **62** | 3 | 7 | | | | |
| **63** | 4 | 9 | | | | |
| **64** | 2 | 4 | | | | |
| **65** | 0 | 2 | | | | |
| **66** | 1 | 2 | | | | |
| **67** | 2 | 5 | | | | |
| **68** | 0 | 2 | | | | |
| **69** | 0 | 3 | | | | |
| **70** | 0 | 2 | | | | |
| **71** | 0 | 0 | | | | |
| **72** | 0 | 0 | | | | |
| **73** | 0 | 0 | | | | |
| **74** | 0 | 0 | | | | |
| **75** | 0 | 0 | | | | |
| **76** | 1 | 1 | | | | |
| **77** | 0 | 1 | | | | |
| **78** | 0 | 3 | | | | |
| **79** | 0 | 3 | | | | |
| **80** | 0 | 0 | | | | |

**Phase I Full-Scale Experiment**
**Test #** 3 **Chamber #** 2
**Pre-ozonated 0.93 μm YG-Microspheres**

| Intensity | Port #1 | Port #2 | Port #3 | Port #4 | Port #5 | Port #b |
|---|---|---|---|---|---|---|
| 1 | | | | 1199 | 1430 | |
| 2 | | | | 3040 | 3116 | |
| 3 | | | | 3419 | 3615 | |
| 4 | | | | 2212 | 2643 | |
| 5 | | | | 935 | 1660 | |
| 6 | | | | 393 | 984 | |
| 7 | | | | 183 | 716 | |
| 8 | | | | 111 | 472 | |
| 9 | | | | 76 | 342 | |
| 10 | | | | 39 | 256 | |
| 11 | | | | 25 | 229 | |
| 12 | | | | 27 | 174 | |
| 13 | | | | 20 | 162 | |
| 14 | | | | 13 | 142 | |
| 15 | | | | 12 | 118 | |
| 16 | | | | 12 | 111 | |
| 17 | | | | 8 | 84 | |
| 18 | | | | 11 | 70 | |
| 19 | | | | 7 | 85 | |
| 20 | | | | 16 | 74 | |
| 21 | | | | 13 | 63 | |
| 22 | | | | 11 | 78 | |
| 23 | | | | 11 | 54 | |
| 24 | | | | 8 | 72 | |
| 25 | | | | 13 | 98 | |
| 26 | | | | 20 | 81 | |
| 27 | | | | 19 | 77 | |
| 28 | | | | 26 | 109 | |
| 29 | | | | 39 | 102 | |
| 30 | | | | 38 | 138 | |
| 31 | | | | 32 | 115 | |
| 32 | | | | 45 | 167 | |
| 33 | | | | 46 | 158 | |
| 34 | | | | 53 | 168 | |
| 35 | | | | 61 | 188 | |
| 36 | | | | 53 | 193 | |
| 37 | | | | 58 | 193 | |
| 38 | | | | 71 | 201 | |
| 39 | | | | 53 | 177 | |
| 40 | | | | 49 | 200 | |

| Intensity | Port #1 | Port #2 | Port #3 | Port #4 | Port #5 | Port #b |
|---|---|---|---|---|---|---|
| 41 | | | | 57 | 172 | |
| 42 | | | | 60 | 191 | |
| 43 | | | | 60 | 182 | |
| 44 | | | | 58 | 145 | |
| 45 | | | | 44 | 140 | |
| 46 | | | | 51 | 133 | |
| 47 | | | | 44 | 125 | |
| 48 | | | | 38 | 106 | |
| 49 | | | | 34 | 67 | |
| 50 | | | | 28 | 67 | |
| 51 | | | | 19 | 60 | |
| 52 | | | | 17 | 53 | |
| 53 | | | | 11 | 46 | |
| 54 | | | | 7 | 25 | |
| 55 | | | | 9 | 28 | |
| 56 | | | | 8 | 14 | |
| 57 | | | | 2 | 12 | |
| 58 | | | | 6 | 18 | |
| 59 | | | | 3 | 10 | |
| 60 | | | | 3 | 10 | |
| 61 | | | | 2 | 9 | |
| 62 | | | | 2 | 3 | |
| 63 | | | | 2 | 6 | |
| 64 | | | | 2 | 5 | |
| 65 | | | | 2 | 2 | |
| 66 | | | | 3 | 4 | |
| 67 | | | | 2 | 1 | |
| 68 | | | | 1 | 3 | |
| 69 | | | | 3 | 1 | |
| 70 | | | | 3 | 3 | |
| 71 | | | | 0 | 4 | |
| 72 | | | | 0 | 0 | |
| 73 | | | | 0 | 2 | |
| 74 | | | | 0 | 1 | |
| 75 | | | | 1 | 1 | |
| 76 | | | | 1 | 0 | |
| 77 | | | | 2 | 3 | |
| 78 | | | | 0 | 1 | |
| 79 | | | | 4 | 0 | |
| 80 | | | | 2 | 2 | |

**Phase I Full-Scale Experiment**
**Test #** 3 **Chamber #** 3,4,5
**Pre-ozonated 0.93 μm YG-Microspheres**

| Chamber | 3 | 4 | 5 |
|---|---|---|---|
| Intensity | Port #5 | Port #5 | Port #5 |
| 1 | 1226 | 5837 | |
| 2 | 2994 | 14450 | |
| 3 | 2874 | 14734 | |
| 4 | 1707 | 7376 | |
| 5 | 731 | 2476 | |
| 6 | 357 | 845 | |
| 7 | 200 | 358 | |
| 8 | 130 | 182 | |
| 9 | 95 | 90 | |
| 10 | 85 | 67 | |
| 11 | 86 | 42 | |
| 12 | 56 | 44 | |
| 13 | 62 | 35 | |
| 14 | 62 | 37 | |
| 15 | 44 | 31 | |
| 16 | 60 | 46 | |
| 17 | 62 | 23 | |
| 18 | 53 | 34 | |
| 19 | 62 | 46 | |
| 20 | 70 | 49 | |
| 21 | 65 | 52 | |
| 22 | 79 | 73 | |
| 23 | 90 | 78 | |
| 24 | 92 | 109 | |
| 25 | 106 | 112 | |
| 26 | 158 | 126 | |
| 27 | 182 | 178 | |
| 28 | 212 | 176 | |
| 29 | 207 | 206 | |
| 30 | 225 | 224 | |
| 31 | 245 | 229 | |
| 32 | 250 | 271 | |
| 33 | 236 | 243 | |
| 34 | 223 | 244 | |
| 35 | 246 | 205 | |
| 36 | 225 | 230 | |
| 37 | 198 | 182 | |
| 38 | 148 | 143 | |
| 39 | 133 | 146 | |
| 40 | 144 | 117 | |

| Chamber | 3 | 4 | 5 |
|---|---|---|---|
| Intensity | Port #5 | Port #5 | Port #5 |
| 41 | 103 | 93 | |
| 42 | 86 | 69 | |
| 43 | 69 | 39 | |
| 44 | 52 | 33 | |
| 45 | 33 | 25 | |
| 46 | 41 | 18 | |
| 47 | 26 | 13 | |
| 48 | 15 | 8 | |
| 49 | 14 | 9 | |
| 50 | 5 | 4 | |
| 51 | 2 | 6 | |
| 52 | 4 | 2 | |
| 53 | 3 | 4 | |
| 54 | 3 | 2 | |
| 55 | 1 | 0 | |
| 56 | 2 | 2 | |
| 57 | 0 | 0 | |
| 58 | 2 | 2 | |
| 59 | 0 | 0 | |
| 60 | 0 | 1 | |
| 61 | 0 | 2 | |
| 62 | 1 | 0 | |
| 63 | 0 | 0 | |
| 64 | 0 | 2 | |
| 65 | 0 | 1 | |
| 66 | 0 | 2 | |
| 67 | 0 | 2 | |
| 68 | 0 | 1 | |
| 69 | 0 | 0 | |
| 70 | 0 | 0 | |
| 71 | 0 | 2 | |
| 72 | 1 | 0 | |
| 73 | 0 | 0 | |
| 74 | 0 | 1 | |
| 75 | 1 | 2 | |
| 76 | 0 | 0 | |
| 77 | 0 | 0 | |
| 78 | 0 | 2 | |
| 79 | 0 | 0 | |
| 80 | 0 | 1 | |

Phase I Full-Scale Experiment
Test # 4 Chamber # 1
Pre-ozonated 0.93 μm YG-Microspheres

| Intensity | Port #1 | Port #2 | Port #3 | Port #4 | Port #5 | Port #b |
|---|---|---|---|---|---|---|
| 1 | 235 | | 185 | 104 | 121 | 312 |
| 2 | 666 | | 474 | 213 | 353 | 536 |
| 3 | 677 | | 426 | 131 | 247 | 299 |
| 4 | 404 | | 215 | 67 | 110 | 84 |
| 5 | 210 | | 95 | 19 | 39 | 15 |
| 6 | 113 | | 48 | 6 | 15 | 8 |
| 7 | 75 | | 29 | 4 | 13 | 6 |
| 8 | 51 | | 20 | 0 | 10 | 4 |
| 9 | 36 | | 10 | 3 | 6 | 3 |
| 10 | 27 | | 17 | 2 | 5 | 6 |
| 11 | 19 | | 11 | 2 | 9 | 8 |
| 12 | 22 | | 6 | 7 | 7 | 5 |
| 13 | 18 | | 8 | 3 | 15 | 8 |
| 14 | 14 | | 4 | 7 | 8 | 14 |
| 15 | 12 | | 2 | 6 | 13 | 13 |
| 16 | 9 | | 7 | 8 | 12 | 21 |
| 17 | 7 | | 7 | 5 | 22 | 14 |
| 18 | 8 | | 6 | 4 | 18 | 15 |
| 19 | 14 | | 15 | 8 | 12 | 20 |
| 20 | 13 | | 13 | 10 | 11 | 27 |
| 21 | 13 | | 17 | 12 | 23 | 22 |
| 22 | 23 | | 13 | 22 | 19 | 38 |
| 23 | 32 | | 19 | 18 | 17 | 24 |
| 24 | 22 | | 24 | 25 | 30 | 31 |
| 25 | 33 | | 17 | 30 | 27 | 32 |
| 26 | 46 | | 35 | 34 | 22 | 53 |
| 27 | 52 | | 33 | 53 | 25 | 49 |
| 28 | 63 | | 35 | 63 | 44 | 43 |
| 29 | 68 | | 60 | 57 | 21 | 53 |
| 30 | 83 | | 45 | 54 | 29 | 49 |
| 31 | 77 | | 59 | 54 | 38 | 49 |
| 32 | 101 | | 70 | 65 | 40 | 45 |
| 33 | 99 | | 66 | 79 | 44 | 60 |
| 34 | 102 | | 80 | 90 | 39 | 56 |
| 35 | 94 | | 80 | 93 | 41 | 54 |
| 36 | 102 | | 91 | 106 | 50 | 52 |
| 37 | 100 | | 77 | 98 | 48 | 59 |
| 38 | 121 | | 92 | 97 | 53 | 72 |
| 39 | 118 | | 83 | 122 | 53 | 43 |
| 40 | 104 | | 112 | 117 | 51 | 74 |

| Intensity | Port #1 | Port #2 | Port #3 | Port #4 | Port #5 | Port #b |
|---|---|---|---|---|---|---|
| 41 | 112 | | 105 | 101 | 45 | 73 |
| 42 | 108 | | 102 | 103 | 59 | 73 |
| 43 | 104 | | 83 | 109 | 57 | 60 |
| 44 | 110 | | 98 | 103 | 48 | 54 |
| 45 | 102 | | 94 | 78 | 57 | 58 |
| 46 | 90 | | 97 | 95 | 50 | 44 |
| 47 | 73 | | 75 | 88 | 48 | 40 |
| 48 | 87 | | 78 | 91 | 49 | 16 |
| 49 | 64 | | 61 | 87 | 41 | 29 |
| 50 | 61 | | 61 | 82 | 35 | 18 |
| 51 | 44 | | 73 | 59 | 39 | 7 |
| 52 | 37 | | 59 | 61 | 26 | 13 |
| 53 | 37 | | 52 | 38 | 24 | 7 |
| 54 | 23 | | 33 | 54 | 22 | 5 |
| 55 | 26 | | 49 | 38 | 17 | 7 |
| 56 | 18 | | 27 | 39 | 25 | 1 |
| 57 | 13 | | 30 | 31 | 15 | 4 |
| 58 | 8 | | 23 | 20 | 15 | 1 |
| 59 | 12 | | 11 | 14 | 8 | 0 |
| 60 | 10 | | 12 | 18 | 7 | 2 |
| 61 | 6 | | 14 | 7 | 7 | 0 |
| 62 | 3 | | 7 | 5 | 6 | 1 |
| 63 | 3 | | 2 | 4 | 2 | 0 |
| 64 | 2 | | 6 | 2 | 2 | 0 |
| 65 | 0 | | 0 | 1 | 0 | 0 |
| 66 | 0 | | 2 | 3 | 0 | 0 |
| 67 | 0 | | 2 | 0 | 0 | 0 |
| 68 | 0 | | 3 | 1 | 0 | 0 |
| 69 | 0 | | 0 | 0 | 2 | 0 |
| 70 | 0 | | 2 | 0 | 0 | 0 |

**Phase I Full-Scale Experiment**
**Test #** 4 **Chamber #** 2
**Pre-ozonated 0.93 μm YG-Microspheres**

| Intensity | Port #1 | Port #2 | Port #3 | Port #4 | Port #5 | Port #b |
|---|---|---|---|---|---|---|
| **1** | 10 | 55 | 105 | 51 | 56 | |
| **2** | 27 | 114 | 205 | 85 | 123 | |
| **3** | 23 | 94 | 108 | 71 | 101 | |
| **4** | 12 | 27 | 39 | 37 | 51 | |
| **5** | 2 | 10 | 15 | 13 | 25 | |
| **6** | 5 | 6 | 4 | 13 | 20 | |
| **7** | 5 | 4 | 9 | 4 | 10 | |
| **8** | 5 | 11 | 5 | 14 | 14 | |
| **9** | 5 | 12 | 5 | 11 | 19 | |
| **10** | 3 | 21 | 15 | 18 | 21 | |
| **11** | 6 | 28 | 16 | 23 | 15 | |
| **12** | 12 | 29 | 17 | 19 | 18 | |
| **13** | 14 | 36 | 21 | 27 | 25 | |
| **14** | 10 | 48 | 22 | 40 | 32 | |
| **15** | 17 | 59 | 40 | 47 | 41 | |
| **16** | 13 | 53 | 52 | 61 | 34 | |
| **17** | 15 | 83 | 58 | 78 | 52 | |
| **18** | 22 | 95 | 59 | 97 | 67 | |
| **19** | 21 | 98 | 73 | 81 | 66 | |
| **20** | 24 | 82 | 77 | 105 | 86 | |
| **21** | 24 | 101 | 86 | 118 | 94 | |
| **22** | 20 | 104 | 83 | 110 | 84 | |
| **23** | 25 | 111 | 108 | 122 | 101 | |
| **24** | 31 | 110 | 110 | 114 | 125 | |
| **25** | 24 | 134 | 92 | 130 | 152 | |
| **26** | 20 | 110 | 118 | 141 | 125 | |
| **27** | 19 | 126 | 109 | 142 | 121 | |
| **28** | 37 | 110 | 146 | 135 | 122 | |
| **29** | 28 | 115 | 136 | 144 | 127 | |
| **30** | 32 | 113 | 113 | 122 | 111 | |
| **31** | 26 | 100 | 104 | 117 | 104 | |
| **32** | 27 | 110 | 100 | 88 | 102 | |
| **33** | 26 | 111 | 122 | 86 | 99 | |
| **34** | 30 | 97 | 94 | 74 | 106 | |
| **35** | 39 | 84 | 82 | 64 | 101 | |
| **36** | 25 | 89 | 54 | 80 | 77 | |
| **37** | 18 | 70 | 76 | 59 | 65 | |
| **38** | 14 | 53 | 68 | 36 | 68 | |
| **39** | 22 | 45 | 60 | 30 | 52 | |
| **40** | 18 | 33 | 55 | 23 | 43 | |

| Intensity | Port #1 | Port #2 | Port #3 | Port #4 | Port #5 | Port #b |
|---|---|---|---|---|---|---|
| **41** | 15 | 24 | 33 | 15 | 36 | |
| **42** | 14 | 17 | 48 | 15 | 40 | |
| **43** | 12 | 16 | 16 | 11 | 34 | |
| **44** | 9 | 12 | 31 | 6 | 14 | |
| **45** | 14 | 7 | 25 | 2 | 15 | |
| **46** | 8 | 6 | 12 | 0 | 8 | |
| **47** | 8 | 13 | 12 | 1 | 6 | |
| **48** | 4 | 4 | 6 | 0 | 5 | |
| **49** | 6 | 2 | 5 | 0 | 4 | |
| **50** | 1 | 0 | 3 | 1 | 2 | |
| **51** | 7 | 1 | 3 | 0 | 4 | |
| **52** | 2 | 1 | 2 | 0 | 0 | |
| **53** | 1 | 0 | 1 | 0 | 4 | |
| **54** | 0 | 0 | 0 | 0 | 3 | |
| **55** | 1 | 0 | 1 | 0 | 0 | |
| **56** | 2 | 0 | 1 | 0 | 1 | |
| **57** | 2 | 0 | 0 | 0 | 0 | |
| **58** | 0 | 0 | 0 | 0 | 1 | |
| **59** | 1 | 0 | 0 | 0 | 0 | |
| **60** | 0 | 0 | 0 | 0 | 0 | |
| **61** | 1 | 0 | 1 | 0 | 0 | |
| **62** | 0 | 0 | 1 | 0 | 0 | |
| **63** | 0 | 0 | 0 | 0 | 0 | |
| **64** | 0 | 0 | 0 | 0 | 0 | |
| **65** | 0 | 0 | 0 | 0 | 0 | |
| **66** | 0 | 0 | 0 | 0 | 0 | |
| **67** | 1 | 0 | 0 | 0 | 0 | |
| **68** | 0 | 0 | 0 | 0 | 0 | |
| **69** | 0 | 0 | 0 | 0 | 0 | |
| **70** | 0 | 0 | 0 | 0 | 0 | |

**Phase I Full-Scale Experiment**
**Test #** 4 **Chamber #** 3
**Pre-ozonated 0.93 μm YG-Microspheres**

| Intensity | Port #1 | Port #2 | Port #3 | Port #4 | Port #5 | Port #b |
|---|---|---|---|---|---|---|
| 1 | 328 | | 426 | 179 | 747 | 150 |
| 2 | 703 | | 735 | 290 | 1369 | 258 |
| 3 | 581 | | 564 | 189 | 927 | 146 |
| 4 | 263 | | 249 | 59 | 330 | 61 |
| 5 | 98 | | 121 | 30 | 123 | 17 |
| 6 | 41 | | 102 | 30 | 52 | 15 |
| 7 | 27 | | 86 | 25 | 38 | 14 |
| 8 | 21 | | 65 | 32 | 43 | 8 |
| 9 | 28 | | 50 | 39 | 58 | 17 |
| 10 | 35 | | 70 | 48 | 43 | 18 |
| 11 | 44 | | 77 | 72 | 53 | 15 |
| 12 | 53 | | 79 | 85 | 52 | 29 |
| 13 | 64 | | 101 | 96 | 73 | 36 |
| 14 | 69 | | 116 | 130 | 81 | 32 |
| 15 | 74 | | 132 | 138 | 78 | 53 |
| 16 | 113 | | 144 | 135 | 120 | 50 |
| 17 | 99 | | 138 | 170 | 128 | 50 |
| 18 | 110 | | 162 | 193 | 108 | 74 |
| 19 | 124 | | 168 | 216 | 116 | 57 |
| 20 | 136 | | 165 | 217 | 123 | 62 |
| 21 | 136 | | 172 | 216 | 173 | 67 |
| 22 | 140 | | 161 | 209 | 163 | 76 |
| 23 | 133 | | 175 | 195 | 138 | 66 |
| 24 | 160 | | 183 | 192 | 140 | 70 |
| 25 | 129 | | 142 | 192 | 167 | 68 |
| 26 | 123 | | 128 | 156 | 149 | 41 |
| 27 | 139 | | 131 | 111 | 147 | 44 |
| 28 | 105 | | 99 | 117 | 106 | 32 |
| 29 | 94 | | 82 | 104 | 107 | 28 |
| 30 | 85 | | 61 | 81 | 81 | 20 |
| 31 | 74 | | 40 | 42 | 87 | 18 |
| 32 | 56 | | 26 | 40 | 61 | 10 |
| 33 | 45 | | 18 | 46 | 53 | 6 |
| 34 | 35 | | 12 | 22 | 47 | 9 |
| 35 | 35 | | 10 | 12 | 37 | 2 |
| 36 | 22 | | 14 | 9 | 30 | 5 |
| 37 | 17 | | 6 | 6 | 18 | 1 |
| 38 | 17 | | 3 | 3 | 15 | 0 |
| 39 | 15 | | 1 | 1 | 16 | 1 |
| 40 | 3 | | 1 | 2 | 5 | 0 |

| Intensity | Port #1 | Port #2 | Port #3 | Port #4 | Port #5 | Port #b |
|---|---|---|---|---|---|---|
| 41 | 6 | | 2 | 1 | 7 | 0 |
| 42 | 8 | | 1 | 0 | 2 | 1 |
| 43 | 5 | | 1 | 0 | 6 | 1 |
| 44 | 1 | | 2 | 0 | 4 | 0 |
| 45 | 4 | | 2 | 0 | 3 | 0 |
| 46 | 3 | | 0 | 0 | 0 | 0 |
| 47 | 0 | | 0 | 0 | 2 | 0 |
| 48 | 1 | | 0 | 0 | 1 | 0 |
| 49 | 1 | | 0 | 0 | 1 | 0 |
| 50 | 0 | | 0 | 0 | 0 | 1 |
| 51 | 0 | | 0 | 0 | 0 | 0 |
| 52 | 0 | | 0 | 0 | 0 | 0 |
| 53 | 0 | | 1 | 0 | 0 | 0 |
| 54 | 0 | | 1 | 0 | 1 | 0 |
| 55 | 1 | | 0 | 0 | 0 | 0 |
| 56 | 0 | | 0 | 0 | 1 | 0 |
| 57 | 0 | | 0 | 0 | 0 | 0 |
| 58 | 0 | | 0 | 0 | 0 | 0 |
| 59 | 0 | | 0 | 1 | 0 | 0 |
| 60 | 0 | | 0 | 0 | 1 | 0 |
| 61 | 0 | | 0 | 0 | 0 | 0 |
| 62 | 0 | | 0 | 0 | 0 | 0 |
| 63 | 0 | | 0 | 0 | 0 | 0 |
| 64 | 0 | | 0 | 0 | 0 | 0 |
| 65 | 1 | | 0 | 0 | 0 | 0 |
| 66 | 0 | | 0 | 0 | 0 | 0 |
| 67 | 0 | | 0 | 0 | 0 | 0 |
| 68 | 0 | | 0 | 0 | 0 | 0 |
| 69 | 0 | | 0 | 0 | 0 | 0 |
| 70 | 0 | | 0 | 0 | 0 | 0 |

**Phase I Full-Scale Experiment**
**Test #** 4 **Chamber #** 4
**Pre-ozonated 0.93 μm YG-Microspheres**

| Intensity | Port #1 | Port #2 | Port #3 | Port #4 | Port #5 | Port #b |
|---|---|---|---|---|---|---|
| **1** | 19 | 52 | | | | 17 |
| **2** | 70 | 107 | | | | 28 |
| **3** | 49 | 89 | | | | 39 |
| **4** | 54 | 61 | | | | 28 |
| **5** | 44 | 41 | | | | 20 |
| **6** | 76 | 45 | | | | 32 |
| **7** | 94 | 66 | | | | 43 |
| **8** | 115 | 84 | | | | 68 |
| **9** | 110 | 109 | | | | 84 |
| **10** | 156 | 132 | | | | 124 |
| **11** | 198 | 143 | | | | 136 |
| **12** | 229 | 165 | | | | 212 |
| **13** | 251 | 177 | | | | 187 |
| **14** | 260 | 186 | | | | 238 |
| **15** | 254 | 224 | | | | 280 |
| **16** | 235 | 246 | | | | 250 |
| **17** | 231 | 269 | | | | 295 |
| **18** | 214 | 257 | | | | 299 |
| **19** | 173 | 265 | | | | 299 |
| **20** | 159 | 211 | | | | 316 |
| **21** | 150 | 196 | | | | 271 |
| **22** | 84 | 182 | | | | 241 |
| **23** | 69 | 160 | | | | 189 |
| **24** | 48 | 133 | | | | 157 |
| **25** | 28 | 120 | | | | 130 |
| **26** | 25 | 91 | | | | 82 |
| **27** | 6 | 67 | | | | 83 |
| **28** | 5 | 45 | | | | 39 |
| **29** | 3 | 35 | | | | 27 |
| **30** | 2 | 25 | | | | 23 |
| **31** | 3 | 13 | | | | 7 |
| **32** | 4 | 14 | | | | 6 |
| **33** | 4 | 9 | | | | 4 |
| **34** | 0 | 3 | | | | 3 |
| **35** | 1 | 1 | | | | 3 |
| **36** | 1 | 1 | | | | 1 |
| **37** | 2 | 1 | | | | 1 |
| **38** | 2 | 0 | | | | 1 |
| **39** | 1 | 0 | | | | 1 |
| **40** | 1 | 0 | | | | 0 |

| Intensity | Port #1 | Port #2 | Port #3 | Port #4 | Port #5 | Port #b |
|---|---|---|---|---|---|---|
| **41** | 4 | 0 | | | | 2 |
| **42** | 1 | 0 | | | | 0 |
| **43** | 0 | 0 | | | | 0 |
| **44** | 0 | 0 | | | | 0 |
| **45** | 1 | 0 | | | | 0 |
| **46** | 0 | 0 | | | | 0 |
| **47** | 1 | 0 | | | | 0 |
| **48** | 0 | 0 | | | | 0 |
| **49** | 0 | 0 | | | | 0 |
| **50** | 0 | 0 | | | | 0 |
| **51** | 0 | 0 | | | | 0 |
| **52** | 0 | 0 | | | | 0 |
| **53** | 0 | 0 | | | | 0 |
| **54** | 0 | 0 | | | | 0 |
| **55** | 0 | 0 | | | | 0 |
| **56** | 0 | 0 | | | | 0 |
| **57** | 0 | 0 | | | | 0 |
| **58** | 0 | 0 | | | | 0 |
| **59** | 0 | 0 | | | | 0 |
| **60** | 0 | 0 | | | | 0 |
| **61** | 0 | 0 | | | | 0 |
| **62** | 0 | 0 | | | | 0 |
| **63** | 0 | 0 | | | | 0 |
| **64** | 0 | 0 | | | | 0 |
| **65** | 0 | 0 | | | | 0 |
| **66** | 0 | 0 | | | | 0 |
| **67** | 0 | 0 | | | | 0 |
| **68** | 0 | 0 | | | | 0 |
| **69** | 0 | 0 | | | | 0 |
| **70** | 1 | 0 | | | | 0 |

**Phase I Full-Scale Experiment**

**Test #** 4 **Chamber #** 5

**Pre-ozonated 0.93 μm YG-Microspheres**

| Intensity | Port #1 | Port #2 | Port #3 | Port #4 | Port #5 | Port #b |
|---|---|---|---|---|---|---|
| 1 | 31 | 781 | 14 | 93 | 141 | 95 |
| 2 | 78 | 1899 | 40 | 181 | 261 | 228 |
| 3 | 115 | 1825 | 54 | 159 | 272 | 175 |
| 4 | 178 | 728 | 19 | 75 | 157 | 110 |
| 5 | 141 | 178 | 27 | 47 | 81 | 64 |
| 6 | 134 | 83 | 34 | 34 | 47 | 59 |
| 7 | 93 | 54 | 53 | 49 | 47 | 59 |
| 8 | 112 | 58 | 69 | 74 | 48 | 79 |
| 9 | 112 | 68 | 76 | 97 | 69 | 89 |
| 10 | 130 | 62 | 95 | 99 | 81 | 119 |
| 11 | 186 | 91 | 118 | 139 | 98 | 130 |
| 12 | 191 | 102 | 133 | 162 | 117 | 161 |
| 13 | 213 | 122 | 142 | 175 | 125 | 211 |
| 14 | 266 | 133 | 141 | 206 | 150 | 217 |
| 15 | 285 | 156 | 177 | 239 | 177 | 220 |
| 16 | 320 | 162 | 171 | 252 | 192 | 240 |
| 17 | 308 | 195 | 175 | 268 | 215 | 211 |
| 18 | 325 | 191 | 147 | 248 | 199 | 232 |
| 19 | 307 | 206 | 157 | 228 | 218 | 204 |
| 20 | 269 | 206 | 137 | 193 | 207 | 177 |
| 21 | 279 | 181 | 102 | 202 | 218 | 141 |
| 22 | 206 | 194 | 115 | 172 | 206 | 136 |
| 23 | 191 | 179 | 51 | 128 | 201 | 107 |
| 24 | 158 | 164 | 54 | 116 | 163 | 97 |
| 25 | 113 | 153 | 43 | 90 | 167 | 65 |
| 26 | 78 | 131 | 25 | 67 | 114 | 41 |
| 27 | 57 | 91 | 27 | 52 | 116 | 35 |
| 28 | 40 | 72 | 14 | 38 | 77 | 31 |
| 29 | 33 | 54 | 9 | 29 | 78 | 8 |
| 30 | 14 | 38 | 6 | 17 | 50 | 19 |
| 31 | 4 | 35 | 5 | 13 | 45 | 6 |
| 32 | 3 | 26 | 4 | 6 | 21 | 4 |
| 33 | 4 | 25 | 3 | 5 | 21 | 2 |
| 34 | 1 | 11 | 3 | 0 | 16 | 3 |
| 35 | 2 | 6 | 2 | 1 | 12 | 6 |
| 36 | 2 | 5 | 1 | 1 | 7 | 1 |
| 37 | 0 | 4 | 2 | 0 | 4 | 2 |
| 38 | 1 | 0 | 3 | 0 | 3 | 1 |
| 39 | 0 | 1 | 1 | 0 | 1 | 0 |
| 40 | 0 | 1 | 0 | 1 | 1 | 2 |

| Intensity | Port #1 | Port #2 | Port #3 | Port #4 | Port #5 | Port #b |
|---|---|---|---|---|---|---|
| 41 | 0 | 1 | 0 | 0 | 1 | 3 |
| 42 | 0 | 0 | 1 | 1 | 0 | 0 |
| 43 | 1 | 0 | 1 | 0 | 1 | 0 |
| 44 | 0 | 0 | 1 | 3 | 0 | 3 |
| 45 | 0 | 0 | 1 | 0 | 0 | 1 |
| 46 | 0 | 0 | 2 | 1 | 0 | 0 |
| 47 | 0 | 0 | 1 | 1 | 0 | 0 |
| 48 | 1 | 0 | 2 | 1 | 0 | 1 |
| 49 | 0 | 0 | 1 | 0 | 0 | 1 |
| 50 | 0 | 0 | 0 | 0 | 1 | 2 |
| 51 | 0 | 0 | 0 | 1 | 0 | 0 |
| 52 | 0 | 1 | 0 | 2 | 0 | 1 |
| 53 | 0 | 0 | 1 | 0 | 0 | 0 |
| 54 | 0 | 0 | 0 | 0 | 0 | 0 |
| 55 | 0 | 0 | 0 | 0 | 0 | 1 |
| 56 | 0 | 0 | 0 | 0 | 0 | 0 |
| 57 | 0 | 0 | 1 | 0 | 1 | 0 |
| 58 | 0 | 0 | 0 | 0 | 0 | 1 |
| 59 | 0 | 0 | 0 | 2 | 0 | 0 |
| 60 | 0 | 0 | 0 | 1 | 0 | 0 |
| 61 | 0 | 0 | 0 | 0 | 1 | 1 |
| 62 | 0 | 0 | 0 | 0 | 0 | 0 |
| 63 | 0 | 0 | 0 | 0 | 0 | 0 |
| 64 | 0 | 0 | 0 | 0 | 0 | 0 |
| 65 | 0 | 0 | 0 | 1 | 0 | 0 |
| 66 | 0 | 0 | 0 | 0 | 0 | 0 |
| 67 | 0 | 0 | 0 | 0 | 0 | 0 |
| 68 | 0 | 0 | 0 | 1 | 0 | 0 |
| 69 | 0 | 0 | 0 | 0 | 0 | 0 |
| 70 | 0 | 0 | 0 | 0 | 0 | 1 |

**Phase I Full-Scale Experiment**
**Test #** 5 **Chamber #** 1
**Pre-ozonated 0.93 μm YG-Microspheres**

| Intensity | Port #1 | Port #2 | Port #3 | Port #4 | Port #5 | Port #b |
|---|---|---|---|---|---|---|
| **1** | 3474 | 1359 | 988 | 80 | 956 | 20 |
| **2** | 8518 | 3422 | 2503 | 211 | 2092 | 39 |
| **3** | 10799 | 3956 | 2897 | 201 | 1785 | 23 |
| **4** | 8391 | 3146 | 1766 | 140 | 750 | 21 |
| **5** | 5531 | 2029 | 768 | 48 | 272 | 15 |
| **6** | 3853 | 1003 | 319 | 28 | 103 | 28 |
| **7** | 2872 | 581 | 154 | 20 | 50 | 8 |
| **8** | 2074 | 296 | 75 | 9 | 30 | 11 |
| **9** | 1663 | 157 | 59 | 13 | 23 | 13 |
| **10** | 1333 | 76 | 31 | 4 | 19 | 10 |
| **11** | 1137 | 47 | 32 | 7 | 14 | 5 |
| **12** | 853 | 30 | 17 | 7 | 6 | 5 |
| **13** | 699 | 27 | 22 | 4 | 12 | 7 |
| **14** | 529 | 29 | 12 | 7 | 7 | 4 |
| **15** | 383 | 6 | 8 | 7 | 6 | 4 |
| **16** | 330 | 9 | 13 | 3 | 11 | 8 |
| **17** | 245 | 8 | 12 | 5 | 3 | 7 |
| **18** | 196 | 17 | 13 | 7 | 2 | 3 |
| **19** | 132 | 4 | 14 | 9 | 6 | 4 |
| **20** | 111 | 10 | 6 | 9 | 7 | 1 |
| **21** | 74 | 6 | 13 | 3 | 8 | 2 |
| **22** | 79 | 13 | 12 | 6 | 3 | 1 |
| **23** | 68 | 15 | 12 | 5 | 4 | 3 |
| **24** | 39 | 9 | 13 | 4 | 6 | 4 |
| **25** | 41 | 11 | 9 | 6 | 11 | 5 |
| **26** | 18 | 13 | 12 | 8 | 6 | 6 |
| **27** | 37 | 9 | 6 | 9 | 11 | 7 |
| **28** | 29 | 12 | 14 | 8 | 9 | 8 |
| **29** | 37 | 16 | 17 | 11 | 5 | 7 |
| **30** | 40 | 8 | 15 | 10 | 7 | 11 |
| **31** | 30 | 18 | 19 | 15 | 9 | 10 |
| **32** | 35 | 23 | 30 | 13 | 18 | 20 |
| **33** | 49 | 26 | 27 | 18 | 17 | 21 |
| **34** | 60 | 27 | 36 | 27 | 19 | 24 |
| **35** | 69 | 44 | 40 | 27 | 19 | 41 |
| **36** | 72 | 46 | 50 | 35 | 34 | 34 |
| **37** | 96 | 57 | 62 | 40 | 43 | 54 |
| **38** | 105 | 54 | 74 | 63 | 46 | 62 |
| **39** | 103 | 82 | 80 | 57 | 44 | 62 |
| **40** | 132 | 66 | 88 | 72 | 70 | 76 |

| Intensity | Port #1 | Port #2 | Port #3 | Port #4 | Port #5 | Port #b |
|---|---|---|---|---|---|---|
| **41** | 146 | 77 | 89 | 87 | 58 | 85 |
| **42** | 144 | 75 | 86 | 95 | 64 | 109 |
| **43** | 154 | 88 | 107 | 104 | 89 | 100 |
| **44** | 175 | 106 | 101 | 97 | 84 | 91 |
| **45** | 155 | 90 | 91 | 124 | 88 | 129 |
| **46** | 162 | 98 | 119 | 135 | 114 | 122 |
| **47** | 189 | 104 | 110 | 124 | 90 | 138 |
| **48** | 173 | 75 | 119 | 140 | 114 | 141 |
| **49** | 172 | 93 | 111 | 151 | 99 | 145 |
| **50** | 184 | 114 | 97 | 109 | 106 | 125 |
| **51** | 155 | 91 | 86 | 116 | 134 | 137 |
| **52** | 169 | 93 | 90 | 144 | 72 | 123 |
| **53** | 143 | 94 | 87 | 116 | 105 | 121 |
| **54** | 112 | 61 | 99 | 106 | 101 | 133 |
| **55** | 113 | 64 | 81 | 94 | 84 | 118 |
| **56** | 87 | 66 | 70 | 121 | 97 | 119 |
| **57** | 94 | 70 | 68 | 98 | 87 | 110 |
| **58** | 70 | 67 | 67 | 89 | 99 | 114 |
| **59** | 60 | 59 | 60 | 109 | 129 | 108 |
| **60** | 55 | 46 | 42 | 82 | 90 | 101 |
| **61** | 29 | 33 | 41 | 94 | 65 | 90 |
| **62** | 38 | 36 | 38 | 80 | 77 | 90 |
| **63** | 31 | 25 | 31 | 54 | 59 | 103 |
| **64** | 34 | 22 | 32 | 60 | 52 | 82 |
| **65** | 20 | 21 | 21 | 53 | 47 | 76 |
| **66** | 14 | 16 | 20 | 49 | 34 | 67 |
| **67** | 10 | 9 | 8 | 33 | 34 | 63 |
| **68** | 7 | 11 | 15 | 24 | 27 | 40 |
| **69** | 6 | 8 | 9 | 21 | 22 | 31 |
| **70** | 3 | 8 | 3 | 19 | 11 | 18 |
| **71** | 6 | 5 | 3 | 16 | 20 | 20 |
| **72** | 4 | 2 | 5 | 12 | 19 | 18 |
| **73** | 4 | 4 | 4 | 8 | 6 | 14 |
| **74** | 2 | 2 | 3 | 6 | 8 | 14 |
| **75** | 2 | 3 | 1 | 8 | 2 | 4 |
| **76** | 6 | 1 | 0 | 4 | 6 | 2 |
| **77** | 5 | 6 | 1 | 3 | 4 | 2 |
| **78** | 3 | 0 | 2 | 1 | 3 | 2 |
| **79** | 0 | 1 | 1 | 1 | 2 | 4 |
| **80** | | | | | | |

**Phase I Full-Scale Experiment**
**Test #** 5 **Chamber #** 2
**Pre-ozonated 0.93 μm YG-Microspheres**

| Intensity | Port #1 | Port #2 | Port #3 | Port #4 | Port #5 | Port #b |
|---|---|---|---|---|---|---|
| 1 | 1122 | 150 | 48 | 150 | 86 | |
| 2 | 2739 | 465 | 116 | 351 | 226 | |
| 3 | 2756 | 469 | 71 | 281 | 225 | |
| 4 | 1147 | 258 | 52 | 123 | 121 | |
| 5 | 304 | 111 | 12 | 66 | 60 | |
| 6 | 91 | 52 | 15 | 41 | 48 | |
| 7 | 38 | 35 | 10 | 28 | 24 | |
| 8 | 18 | 19 | 9 | 21 | 14 | |
| 9 | 12 | 15 | 4 | 16 | 18 | |
| 10 | 10 | 9 | 7 | 8 | 11 | |
| 11 | 7 | 9 | 3 | 12 | 7 | |
| 12 | 9 | 8 | 4 | 2 | 9 | |
| 13 | 3 | 5 | 5 | 8 | 12 | |
| 14 | 3 | 1 | 7 | 4 | 6 | |
| 15 | 15 | 5 | 2 | 13 | 7 | |
| 16 | 10 | 4 | 7 | 12 | 2 | |
| 17 | 12 | 1 | 4 | 7 | 1 | |
| 18 | 12 | 5 | 9 | 8 | 10 | |
| 19 | 13 | 3 | 12 | 10 | 8 | |
| 20 | 18 | 6 | 8 | 13 | 19 | |
| 21 | 9 | 10 | 10 | 13 | 15 | |
| 22 | 15 | 8 | 26 | 13 | 24 | |
| 23 | 10 | 9 | 21 | 21 | 22 | |
| 24 | 17 | 12 | 36 | 21 | 37 | |
| 25 | 18 | 21 | 40 | 33 | 60 | |
| 26 | 22 | 20 | 48 | 44 | 49 | |
| 27 | 17 | 13 | 78 | 46 | 91 | |
| 28 | 32 | 22 | 76 | 71 | 81 | |
| 29 | 34 | 28 | 99 | 93 | 98 | |
| 30 | 39 | 34 | 103 | 88 | 118 | |
| 31 | 55 | 30 | 145 | 114 | 156 | |
| 32 | 56 | 48 | 162 | 118 | 150 | |
| 33 | 60 | 65 | 178 | 145 | 154 | |
| 34 | 74 | 65 | 174 | 169 | 163 | |
| 35 | 100 | 76 | 178 | 166 | 183 | |
| 36 | 107 | 98 | 185 | 199 | 168 | |
| 37 | 117 | 106 | 221 | 205 | 187 | |
| 38 | 124 | 111 | 222 | 243 | 145 | |
| 39 | 144 | 139 | 204 | 187 | 155 | |
| 40 | 149 | 140 | 203 | 212 | 157 | |

| Intensity | Port #1 | Port #2 | Port #3 | Port #4 | Port #5 | Port #b |
|---|---|---|---|---|---|---|
| 41 | 152 | 169 | 190 | 203 | 161 | |
| 42 | 159 | 153 | 163 | 193 | 143 | |
| 43 | 155 | 161 | 154 | 161 | 120 | |
| 44 | 155 | 155 | 140 | 178 | 105 | |
| 45 | 153 | 192 | 126 | 139 | 104 | |
| 46 | 139 | 186 | 105 | 122 | 76 | |
| 47 | 149 | 146 | 107 | 103 | 67 | |
| 48 | 132 | 152 | 87 | 90 | 56 | |
| 49 | 124 | 147 | 74 | 83 | 53 | |
| 50 | 129 | 158 | 58 | 38 | 58 | |
| 51 | 110 | 156 | 39 | 44 | 39 | |
| 52 | 108 | 137 | 38 | 31 | 34 | |
| 53 | 84 | 106 | 29 | 26 | 27 | |
| 54 | 89 | 105 | 20 | 25 | 28 | |
| 55 | 77 | 92 | 11 | 15 | 27 | |
| 56 | 72 | 77 | 16 | 13 | 17 | |
| 57 | 57 | 76 | 7 | 4 | 18 | |
| 58 | 56 | 72 | 5 | 5 | 11 | |
| 59 | 33 | 43 | 7 | 4 | 14 | |
| 60 | 32 | 43 | 5 | 2 | 13 | |
| 61 | 32 | 36 | 2 | 1 | 6 | |
| 62 | 19 | 30 | 1 | 1 | 6 | |
| 63 | 23 | 26 | 1 | 2 | 8 | |
| 64 | 12 | 14 | 3 | 1 | 5 | |
| 65 | 9 | 13 | 1 | 0 | 6 | |
| 66 | 8 | 13 | 1 | 1 | 5 | |
| 67 | 9 | 15 | 0 | 0 | 3 | |
| 68 | 5 | 7 | 0 | 0 | 2 | |
| 69 | 2 | 4 | 0 | 1 | 2 | |
| 70 | 1 | 3 | 0 | 1 | 0 | |
| 71 | 3 | 8 | 0 | 0 | 0 | |
| 72 | 4 | 3 | 0 | 0 | 0 | |
| 73 | 3 | 1 | 0 | 0 | 2 | |
| 74 | 1 | 0 | 0 | 0 | 0 | |
| 75 | 1 | 2 | 0 | 0 | 1 | |
| 76 | 0 | 1 | 0 | 0 | 0 | |
| 77 | 0 | 1 | 0 | 0 | 1 | |
| 78 | 1 | 0 | 0 | 0 | 0 | |
| 79 | 1 | 0 | 0 | 0 | 1 | |
| 80 | 1 | 0 | 0 | 0 | 0 | |

Phase I Full-Scale Experiment

Test # 5 Chamber # 4

Pre-ozonated 0.93 μm YG-Microspheres

| Intensity | Port #1 | Port #2 | Port #3 | Port #4 | Port #5 | Port #b |
|---|---|---|---|---|---|---|
| 1 | 66 | | 24 | | 155 | |
| 2 | 193 | | 73 | | 341 | |
| 3 | 182 | | 54 | | 360 | |
| 4 | 112 | | 29 | | 183 | |
| 5 | 87 | | 30 | | 83 | |
| 6 | 49 | | 9 | | 29 | |
| 7 | 39 | | 10 | | 18 | |
| 8 | 37 | | 9 | | 16 | |
| 9 | 31 | | 4 | | 11 | |
| 10 | 32 | | 7 | | 9 | |
| 11 | 17 | | 9 | | 8 | |
| 12 | 17 | | 6 | | 8 | |
| 13 | 16 | | 6 | | 9 | |
| 14 | 15 | | 2 | | 8 | |
| 15 | 13 | | 5 | | 9 | |
| 16 | 10 | | 8 | | 11 | |
| 17 | 14 | | 5 | | 15 | |
| 18 | 12 | | 9 | | 18 | |
| 19 | 13 | | 15 | | 29 | |
| 20 | 19 | | 21 | | 20 | |
| 21 | 20 | | 21 | | 34 | |
| 22 | 27 | | 36 | | 48 | |
| 23 | 25 | | 50 | | 47 | |
| 24 | 35 | | 59 | | 76 | |
| 25 | 40 | | 75 | | 91 | |
| 26 | 64 | | 84 | | 124 | |
| 27 | 66 | | 115 | | 150 | |
| 28 | 90 | | 127 | | 198 | |
| 29 | 111 | | 189 | | 205 | |
| 30 | 128 | | 176 | | 222 | |
| 31 | 146 | | 194 | | 257 | |
| 32 | 155 | | 205 | | 243 | |
| 33 | 175 | | 223 | | 264 | |
| 34 | 201 | | 251 | | 260 | |
| 35 | 207 | | 253 | | 250 | |
| 36 | 195 | | 224 | | 222 | |
| 37 | 208 | | 236 | | 216 | |
| 38 | 196 | | 184 | | 192 | |
| 39 | 198 | | 186 | | 138 | |
| 40 | 153 | | 168 | | 123 | |

| Intensity | Port #1 | Port #2 | Port #3 | Port #4 | Port #5 | Port #b |
|---|---|---|---|---|---|---|
| 41 | 162 | | 137 | | 104 | |
| 42 | 164 | | 107 | | 77 | |
| 43 | 95 | | 76 | | 54 | |
| 44 | 85 | | 57 | | 39 | |
| 45 | 88 | | 37 | | 25 | |
| 46 | 51 | | 33 | | 20 | |
| 47 | 47 | | 32 | | 18 | |
| 48 | 27 | | 10 | | 13 | |
| 49 | 22 | | 5 | | 6 | |
| 50 | 16 | | 2 | | 4 | |
| 51 | 15 | | 6 | | 6 | |
| 52 | 3 | | 3 | | 0 | |
| 53 | 3 | | 1 | | 3 | |
| 54 | 4 | | 4 | | 3 | |
| 55 | 2 | | 1 | | 0 | |
| 56 | 2 | | 1 | | 0 | |
| 57 | 2 | | 2 | | 0 | |
| 58 | 2 | | 0 | | 0 | |
| 59 | 1 | | 0 | | 0 | |
| 60 | 2 | | 0 | | 0 | |
| 61 | 1 | | 0 | | 0 | |
| 62 | 1 | | 1 | | 0 | |
| 63 | 0 | | 0 | | 1 | |
| 64 | 0 | | 0 | | 0 | |
| 65 | 0 | | 0 | | 1 | |
| 66 | 1 | | 1 | | 0 | |
| 67 | 0 | | 0 | | 1 | |
| 68 | 2 | | 0 | | 0 | |
| 69 | 0 | | 0 | | 0 | |
| 70 | 0 | | 0 | | 0 | |
| 71 | 0 | | 0 | | 0 | |
| 72 | 0 | | 0 | | 0 | |
| 73 | 0 | | 0 | | 1 | |
| 74 | 0 | | 0 | | 0 | |
| 75 | 0 | | 0 | | 0 | |
| 76 | 0 | | 0 | | 0 | |
| 77 | 0 | | 0 | | 0 | |
| 78 | 0 | | 0 | | 0 | |
| 79 | 0 | | 0 | | 1 | |
| 80 | 0 | | 0 | | 1 | |

**Phase I Full-Scale Experiment**
**Test #** 5 **Chamber #** 5
**Pre-ozonated 0.93 μm YG-Microspheres**

| Intensity | Port #1 | Port #2 | Port #3 | Port #4 | Port #5 | Port #b |
|---|---|---|---|---|---|---|
| **1** | 96 | | 285 | | 53 | |
| **2** | 180 | | 1381 | | 92 | |
| **3** | 153 | | 3493 | | 107 | |
| **4** | 65 | | 4063 | | 48 | |
| **5** | 27 | | 2911 | | 37 | |
| **6** | 18 | | 1717 | | 20 | |
| **7** | 11 | | 928 | | 16 | |
| **8** | 12 | | 607 | | 12 | |
| **9** | 11 | | 363 | | 8 | |
| **10** | 10 | | 275 | | 5 | |
| **11** | 7 | | 179 | | 5 | |
| **12** | 4 | | 126 | | 2 | |
| **13** | 6 | | 85 | | 1 | |
| **14** | 3 | | 64 | | 4 | |
| **15** | 6 | | 46 | | 7 | |
| **16** | 14 | | 42 | | 3 | |
| **17** | 8 | | 27 | | 2 | |
| **18** | 13 | | 24 | | 6 | |
| **19** | 9 | | 17 | | 12 | |
| **20** | 7 | | 15 | | 14 | |
| **21** | 13 | | 34 | | 11 | |
| **22** | 19 | | 39 | | 19 | |
| **23** | 30 | | 45 | | 23 | |
| **24** | 27 | | 60 | | 33 | |
| **25** | 42 | | 66 | | 33 | |
| **26** | 38 | | 105 | | 46 | |
| **27** | 63 | | 133 | | 77 | |
| **28** | 80 | | 152 | | 88 | |
| **29** | 100 | | 193 | | 105 | |
| **30** | 124 | | 207 | | 118 | |
| **31** | 147 | | 248 | | 130 | |
| **32** | 202 | | 237 | | 170 | |
| **33** | 209 | | 309 | | 204 | |
| **34** | 229 | | 304 | | 218 | |
| **35** | 251 | | 292 | | 239 | |
| **36** | 241 | | 248 | | 264 | |
| **37** | 226 | | 243 | | 280 | |
| **38** | 225 | | 186 | | 262 | |
| **39** | 217 | | 190 | | 268 | |
| **40** | 234 | | 167 | | 251 | |

| Intensity | Port #1 | Port #2 | Port #3 | Port #4 | Port #5 | Port #b |
|---|---|---|---|---|---|---|
| **41** | 187 | | 120 | | 192 | |
| **42** | 133 | | 84 | | 208 | |
| **43** | 121 | | 72 | | 169 | |
| **44** | 77 | | 52 | | 163 | |
| **45** | 69 | | 41 | | 107 | |
| **46** | 64 | | 34 | | 87 | |
| **47** | 44 | | 17 | | 70 | |
| **48** | 25 | | 16 | | 52 | |
| **49** | 13 | | 12 | | 27 | |
| **50** | 6 | | 4 | | 21 | |
| **51** | 7 | | 3 | | 20 | |
| **52** | 7 | | 2 | | 8 | |
| **53** | 1 | | 1 | | 9 | |
| **54** | 2 | | 1 | | 6 | |
| **55** | 0 | | 1 | | 3 | |
| **56** | 0 | | 0 | | 2 | |
| **57** | 1 | | 0 | | 0 | |
| **58** | 0 | | 1 | | 0 | |
| **59** | 0 | | 0 | | 1 | |
| **60** | 0 | | 0 | | 0 | |
| **61** | 0 | | 0 | | 0 | |
| **62** | 0 | | 0 | | 0 | |
| **63** | 0 | | 1 | | 0 | |
| **64** | 0 | | 0 | | 0 | |
| **65** | 0 | | 0 | | 0 | |
| **66** | 0 | | 0 | | 0 | |
| **67** | 0 | | 0 | | 0 | |
| **68** | 0 | | 0 | | 0 | |
| **69** | 0 | | 0 | | 0 | |
| **70** | 0 | | 0 | | 0 | |
| **71** | 0 | | 0 | | 0 | |
| **72** | 0 | | 0 | | 0 | |
| **73** | 0 | | 0 | | 0 | |
| **74** | 0 | | 0 | | 0 | |
| **75** | 0 | | 0 | | 0 | |
| **76** | 0 | | 0 | | 0 | |
| **77** | 0 | | 0 | | 0 | |
| **78** | 0 | | 0 | | 0 | |
| **79** | 0 | | 0 | | 0 | |
| **80** | 0 | | 0 | | 0 | |

**Phase I Full-Scale Experiment**
**Test #** 6 **Chamber #** 1
**Pre-ozonated 0.93 μm YG-Microspheres**

| Intensity | Port #1 | Port #2 | Port #3 | Port #4 | Port #5 | Port #b |
|---|---|---|---|---|---|---|
| **1** | | | 55 | 236 | 175 | |
| **2** | | | 133 | 540 | 350 | |
| **3** | | | 124 | 556 | 254 | |
| **4** | | | 85 | 372 | 94 | |
| **5** | | | 38 | 313 | 37 | |
| **6** | | | 16 | 247 | 28 | |
| **7** | | | 11 | 205 | 16 | |
| **8** | | | 15 | 142 | 11 | |
| **9** | | | 6 | 107 | 9 | |
| **10** | | | 6 | 60 | 10 | |
| **11** | | | 8 | 65 | 7 | |
| **12** | | | 4 | 46 | 10 | |
| **13** | | | 7 | 39 | 8 | |
| **14** | | | 3 | 31 | 4 | |
| **15** | | | 5 | 33 | 2 | |
| **16** | | | 5 | 38 | 9 | |
| **17** | | | 8 | 27 | 12 | |
| **18** | | | 7 | 27 | 7 | |
| **19** | | | 8 | 33 | 1 | |
| **20** | | | 9 | 24 | 9 | |
| **21** | | | 12 | 24 | 17 | |
| **22** | | | 10 | 25 | 12 | |
| **23** | | | 11 | 17 | 29 | |
| **24** | | | 15 | 14 | 20 | |
| **25** | | | 12 | 17 | 17 | |
| **26** | | | 16 | 22 | 33 | |
| **27** | | | 26 | 29 | 35 | |
| **28** | | | 27 | 30 | 43 | |
| **29** | | | 32 | 36 | 56 | |
| **30** | | | 39 | 48 | 83 | |
| **31** | | | 53 | 43 | 101 | |
| **32** | | | 58 | 51 | 112 | |
| **33** | | | 61 | 62 | 135 | |
| **34** | | | 72 | 67 | 114 | |
| **35** | | | 89 | 75 | 146 | |
| **36** | | | 77 | 90 | 130 | |
| **37** | | | 107 | 98 | 157 | |
| **38** | | | 103 | 97 | 141 | |
| **39** | | | 106 | 109 | 154 | |
| **40** | | | 113 | 90 | 144 | |

| Intensity | Port #1 | Port #2 | Port #3 | Port #4 | Port #5 | Port #b |
|---|---|---|---|---|---|---|
| **41** | | | 110 | 100 | 161 | |
| **42** | | | 124 | 117 | 119 | |
| **43** | | | 146 | 113 | 127 | |
| **44** | | | 115 | 129 | 115 | |
| **45** | | | 117 | 117 | 120 | |
| **46** | | | 126 | 103 | 94 | |
| **47** | | | 121 | 101 | 82 | |
| **48** | | | 97 | 102 | 68 | |
| **49** | | | 94 | 93 | 54 | |
| **50** | | | 85 | 95 | 50 | |
| **51** | | | 105 | 82 | 36 | |
| **52** | | | 81 | 83 | 31 | |
| **53** | | | 75 | 85 | 23 | |
| **54** | | | 55 | 77 | 20 | |
| **55** | | | 47 | 62 | 13 | |
| **56** | | | 50 | 59 | 16 | |
| **57** | | | 40 | 39 | 13 | |
| **58** | | | 31 | 42 | 8 | |
| **59** | | | 22 | 30 | 4 | |
| **60** | | | 17 | 18 | 1 | |
| **61** | | | 20 | 17 | 0 | |
| **62** | | | 10 | 19 | 1 | |
| **63** | | | 7 | 9 | 0 | |
| **64** | | | 5 | 8 | 0 | |
| **65** | | | 2 | 6 | 1 | |
| **66** | | | 2 | 4 | 0 | |
| **67** | | | 0 | 5 | 0 | |
| **68** | | | 2 | 2 | 0 | |
| **69** | | | 3 | 0 | 1 | |
| **70** | | | 1 | 1 | 0 | |
| **71** | | | 2 | 0 | 0 | |
| **72** | | | 0 | 0 | 1 | |
| **73** | | | 0 | 0 | 0 | |
| **74** | | | 1 | 1 | 0 | |
| **75** | | | 1 | 0 | 0 | |
| **76** | | | 1 | 0 | 0 | |
| **77** | | | 0 | 0 | 0 | |
| **78** | | | 0 | 0 | 0 | |
| **79** | | | 0 | 2 | 0 | |
| **80** | | | 0 | 0 | 0 | |

**Phase I Full-Scale Experiment**
**Test #** 6 **Chamber #** 2
**Pre-ozonated 0.93 μm YG-Microspheres**

| Intensity | Port #1 | Port #2 | Port #3 | Port #4 | Port #5 | Port #b |
|---|---|---|---|---|---|---|
| **1** | | 86 | 175 | 321 | | |
| **2** | | 221 | 350 | 719 | | |
| **3** | | 320 | 254 | 609 | | |
| **4** | | 339 | 94 | 319 | | |
| **5** | | 398 | 37 | 209 | | |
| **6** | | 381 | 28 | 181 | | |
| **7** | | 229 | 16 | 152 | | |
| **8** | | 104 | 11 | 96 | | |
| **9** | | 50 | 9 | 65 | | |
| **10** | | 30 | 10 | 52 | | |
| **11** | | 13 | 7 | 34 | | |
| **12** | | 8 | 10 | 27 | | |
| **13** | | 15 | 8 | 28 | | |
| **14** | | 10 | 4 | 31 | | |
| **15** | | 10 | 2 | 29 | | |
| **16** | | 11 | 9 | 30 | | |
| **17** | | 16 | 12 | 47 | | |
| **18** | | 18 | 7 | 48 | | |
| **19** | | 14 | 1 | 43 | | |
| **20** | | 30 | 9 | 61 | | |
| **21** | | 18 | 17 | 63 | | |
| **22** | | 25 | 12 | 67 | | |
| **23** | | 40 | 29 | 53 | | |
| **24** | | 56 | 20 | 95 | | |
| **25** | | 61 | 17 | 87 | | |
| **26** | | 65 | 33 | 118 | | |
| **27** | | 67 | 35 | 121 | | |
| **28** | | 78 | 43 | 113 | | |
| **29** | | 77 | 56 | 148 | | |
| **30** | | 92 | 83 | 158 | | |
| **31** | | 99 | 101 | 145 | | |
| **32** | | 125 | 112 | 142 | | |
| **33** | | 140 | 135 | 162 | | |
| **34** | | 118 | 114 | 152 | | |
| **35** | | 124 | 146 | 146 | | |
| **36** | | 134 | 130 | 136 | | |
| **37** | | 122 | 157 | 94 | | |
| **38** | | 128 | 141 | 113 | | |
| **39** | | 126 | 154 | 96 | | |
| **40** | | 103 | 144 | 61 | | |

| Intensity | Port #1 | Port #2 | Port #3 | Port #4 | Port #5 | Port #b |
|---|---|---|---|---|---|---|
| **41** | | 106 | 161 | 70 | | |
| **42** | | 122 | 119 | 70 | | |
| **43** | | 90 | 127 | 48 | | |
| **44** | | 94 | 115 | 42 | | |
| **45** | | 87 | 120 | 38 | | |
| **46** | | 76 | 94 | 28 | | |
| **47** | | 60 | 82 | 24 | | |
| **48** | | 64 | 68 | 22 | | |
| **49** | | 44 | 54 | 18 | | |
| **50** | | 43 | 50 | 18 | | |
| **51** | | 29 | 36 | 7 | | |
| **52** | | 35 | 31 | 4 | | |
| **53** | | 15 | 23 | 8 | | |
| **54** | | 18 | 20 | 6 | | |
| **55** | | 12 | 13 | 0 | | |
| **56** | | 9 | 16 | 0 | | |
| **57** | | 11 | 13 | 0 | | |
| **58** | | 7 | 8 | 1 | | |
| **59** | | 3 | 4 | 2 | | |
| **60** | | 6 | 1 | 0 | | |
| **61** | | 3 | 0 | 1 | | |
| **62** | | 0 | 1 | 0 | | |
| **63** | | 1 | 0 | 0 | | |
| **64** | | 0 | 0 | 0 | | |
| **65** | | 1 | 1 | 0 | | |
| **66** | | 0 | 0 | 0 | | |
| **67** | | 0 | 0 | 1 | | |
| **68** | | 0 | 0 | 0 | | |
| **69** | | 0 | 1 | 0 | | |
| **70** | | 2 | 0 | 1 | | |
| **71** | | 2 | 0 | 0 | | |
| **72** | | 0 | 1 | 0 | | |
| **73** | | 0 | 0 | 0 | | |
| **74** | | 0 | 0 | 0 | | |
| **75** | | 0 | 0 | 1 | | |
| **76** | | 0 | 0 | 0 | | |
| **77** | | 0 | 0 | 0 | | |
| **78** | | 0 | 0 | 0 | | |
| **79** | | 0 | 0 | 0 | | |
| **80** | | 0 | 0 | 0 | | |

**Phase I Full-Scale Experiment**

**Test #** 6 **Chamber #** 3,4,5

**Pre-ozonated 0.93 μm YG-Microspheres**

| Chamber | 3 | 4 | 5 |
|---|---|---|---|
| Intensity | Port #5 | Port #5 | Port #5 |
| 1 | 53 | 29 | 15 |
| 2 | 122 | 92 | 53 |
| 3 | 134 | 150 | 107 |
| 4 | 80 | 118 | 94 |
| 5 | 54 | 45 | 30 |
| 6 | 37 | 23 | 9 |
| 7 | 24 | 18 | 6 |
| 8 | 14 | 20 | 6 |
| 9 | 11 | 10 | 6 |
| 10 | 14 | 4 | 4 |
| 11 | 11 | 2 | 8 |
| 12 | 11 | 8 | 8 |
| 13 | 11 | 3 | 25 |
| 14 | 25 | 7 | 16 |
| 15 | 18 | 7 | 28 |
| 16 | 30 | 16 | 27 |
| 17 | 40 | 16 | 37 |
| 18 | 45 | 19 | 48 |
| 19 | 54 | 26 | 61 |
| 20 | 83 | 30 | 91 |
| 21 | 82 | 37 | 115 |
| 22 | 111 | 45 | 149 |
| 23 | 121 | 70 | 179 |
| 24 | 150 | 77 | 167 |
| 25 | 181 | 62 | 181 |
| 26 | 150 | 74 | 203 |
| 27 | 188 | 66 | 204 |
| 28 | 195 | 85 | 198 |
| 29 | 185 | 72 | 195 |
| 30 | 181 | 63 | 169 |
| 31 | 175 | 66 | 148 |
| 32 | 138 | 39 | 117 |
| 33 | 100 | 34 | 106 |
| 34 | 87 | 25 | 85 |
| 35 | 83 | 18 | 57 |
| 36 | 73 | 18 | 47 |
| 37 | 43 | 7 | 35 |
| 38 | 35 | 5 | 18 |
| 39 | 33 | 7 | 14 |
| 40 | 19 | 1 | 10 |

| Chamber | 3 | 4 | 5 |
|---|---|---|---|
| Intensity | Port #5 | Port #5 | Port #5 |
| 41 | 16 | 0 | 2 |
| 42 | 4 | 2 | 4 |
| 43 | 3 | 0 | 0 |
| 44 | 4 | 0 | 2 |
| 45 | 2 | 0 | 2 |
| 46 | 0 | 1 | 0 |
| 47 | 0 | 0 | 1 |
| 48 | 3 | 2 | 1 |
| 49 | 1 | 0 | 1 |
| 50 | 1 | 0 | 0 |
| 51 | 0 | 0 | 0 |
| 52 | 0 | 0 | 0 |
| 53 | 1 | 2 | 0 |
| 54 | 0 | 0 | 0 |
| 55 | 2 | 0 | 0 |
| 56 | 0 | 0 | 0 |
| 57 | 0 | 0 | 0 |
| 58 | 0 | 0 | 0 |
| 59 | 0 | 0 | 0 |
| 60 | 0 | 0 | 0 |
| 61 | 0 | 0 | 0 |
| 62 | 0 | 0 | 0 |
| 63 | 0 | 0 | 0 |
| 64 | 0 | 0 | 0 |
| 65 | 0 | 0 | 0 |
| 66 | 0 | 0 | 0 |
| 67 | 0 | 0 | 0 |
| 68 | 0 | 0 | 0 |
| 69 | 0 | 0 | 0 |
| 70 | 0 | 0 | 0 |
| 71 | 1 | 0 | 1 |
| 72 | 0 | 0 | 0 |
| 73 | 0 | 0 | 0 |
| 74 | 0 | 0 | 0 |
| 75 | 0 | 0 | 0 |
| 76 | 1 | 0 | 0 |
| 77 | 1 | 0 | 0 |
| 78 | 0 | 0 | 0 |
| 79 | 1 | 0 | 0 |
| 80 | 0 | 0 | 0 |

**Phase II Semi-Batch Experiment**
**Pre-ozonated 0.94 μm YG-Microspheres**

| | |
|---|---|
| **Temp.** | 20 °C |
| **pH** | 7 |
| **$[O_3]$** | 0.906 mg/L |

| | *CT* (mg-min/L) | | | | | |
|---|---|---|---|---|---|---|
| **Intensity** | **0** | **1** | **2** | **3** | **4** | **5** |
| **1** | 0 | 0 | 0 | 0 | 0 | 0 |
| **2** | 0 | 0 | 0 | 0 | 0 | 0 |
| **3** | 0 | 0 | 0 | 0 | 0 | 0 |
| **4** | 0 | 0 | 0 | 1 | 0 | 0 |
| **5** | 0 | 0 | 0 | 2 | 1 | 0 |
| **6** | 0 | 0 | 0 | 3 | 0 | 1 |
| **7** | 0 | 1 | 1 | 1 | 1 | 3 |
| **8** | 0 | 4 | 0 | 6 | 4 | 7 |
| **9** | 0 | 4 | 2 | 1 | 4 | 8 |
| **10** | 0 | 3 | 4 | 4 | 11 | 14 |
| **11** | 0 | 1 | 3 | 6 | 11 | 15 |
| **12** | 0 | 5 | 2 | 10 | 7 | 25 |
| **13** | 0 | 4 | 12 | 9 | 17 | 33 |
| **14** | 0 | 5 | 10 | 9 | 17 | 36 |
| **15** | 0 | 5 | 10 | 18 | 35 | 62 |
| **16** | 0 | 6 | 16 | 18 | 53 | 72 |
| **17** | 2 | 7 | 14 | 35 | 48 | 104 |
| **18** | 1 | 9 | 25 | 43 | 87 | 109 |
| **19** | 1 | 16 | 25 | 48 | 71 | 153 |
| **20** | 0 | 12 | 31 | 65 | 110 | 187 |
| **21** | 1 | 12 | 45 | 87 | 113 | 187 |
| **22** | 3 | 19 | 46 | 97 | 174 | 283 |
| **23** | 3 | 29 | 64 | 126 | 193 | 279 |
| **24** | 2 | 34 | 81 | 131 | 208 | 282 |
| **25** | 3 | 31 | 85 | 173 | 264 | 311 |
| **26** | 4 | 50 | 106 | 181 | 257 | 314 |
| **27** | 2 | 65 | 128 | 211 | 287 | 308 |
| **28** | 1 | 67 | 151 | 235 | 285 | 298 |
| **29** | 4 | 80 | 172 | 223 | 286 | 328 |
| **30** | 4 | 76 | 186 | 297 | 281 | 293 |
| **31** | 11 | 107 | 220 | 273 | 298 | 284 |
| **32** | 6 | 106 | 249 | 246 | 254 | 253 |
| **33** | 6 | 141 | 240 | 287 | 267 | 216 |
| **34** | 10 | 174 | 233 | 265 | 247 | 205 |
| **35** | 10 | 175 | 260 | 267 | 247 | 168 |
| **36** | 17 | 173 | 216 | 259 | 212 | 153 |
| **37** | 10 | 210 | 233 | 217 | 181 | 138 |
| **38** | 13 | 195 | 235 | 216 | 160 | 112 |
| **39** | 17 | 207 | 227 | 204 | 142 | 90 |
| **40** | 14 | 224 | 206 | 187 | 136 | 50 |
| **41** | 21 | 217 | 206 | 162 | 112 | 39 |
| **42** | 12 | 206 | 185 | 158 | 79 | 31 |
| **43** | 28 | 176 | 182 | 136 | 75 | 27 |
| **44** | 26 | 178 | 203 | 118 | 53 | 14 |
| **45** | 43 | 183 | 159 | 93 | 42 | 7 |
| **46** | 35 | 176 | 143 | 96 | 34 | 6 |
| **47** | 51 | 201 | 147 | 67 | 14 | 1 |
| **48** | 53 | 166 | 111 | 35 | 16 | 1 |
| **49** | 61 | 196 | 115 | 41 | 9 | 2 |
| **50** | 58 | 165 | 93 | 28 | 11 | 5 |
| **51** | 59 | 157 | 81 | 14 | 5 | 0 |
| **52** | 78 | 149 | 67 | 16 | 2 | 1 |
| **53** | 75 | 145 | 49 | 9 | 2 | 0 |
| **54** | 96 | 117 | 35 | 8 | 0 | 0 |
| **55** | 83 | 123 | 23 | 1 | 1 | 0 |
| **56** | 106 | 98 | 25 | 6 | 2 | 0 |
| **57** | 119 | 89 | 18 | 2 | 0 | 0 |
| **58** | 106 | 93 | 13 | 2 | 0 | 0 |
| **59** | 134 | 55 | 6 | 1 | 0 | 0 |
| **60** | 143 | 51 | 4 | 2 | 0 | 0 |

| | *CT* (mg-min/L) | | | | | |
|---|---|---|---|---|---|---|
| **Intensity** | **0** | **1** | **2** | **3** | **4** | **5** |
| **61** | 151 | 45 | 5 | 0 | 0 | 0 |
| **62** | 144 | 42 | 4 | 0 | 0 | 0 |
| **63** | 157 | 34 | 0 | 0 | 0 | 0 |
| **64** | 146 | 29 | 1 | 0 | 0 | 0 |
| **65** | 137 | 14 | 0 | 0 | 0 | 0 |
| **66** | 157 | 12 | 0 | 0 | 0 | 0 |
| **67** | 177 | 8 | 0 | 0 | 0 | 0 |
| **68** | 129 | 9 | 0 | 0 | 0 | 0 |
| **69** | 157 | 7 | 0 | 0 | 0 | 0 |
| **70** | 150 | 4 | 0 | 0 | 0 | 0 |
| **71** | 124 | 2 | 0 | 0 | 0 | 0 |
| **72** | 148 | 1 | 0 | 0 | 0 | 0 |
| **73** | 144 | 3 | 0 | 0 | 0 | 0 |
| **74** | 138 | 2 | 0 | 0 | 0 | 0 |
| **75** | 128 | 0 | 0 | 0 | 0 | 0 |
| **76** | 143 | 0 | 0 | 0 | 0 | 0 |
| **77** | 124 | 0 | 0 | 0 | 0 | 0 |
| **78** | 140 | 0 | 0 | 0 | 0 | 0 |
| **79** | 132 | 0 | 0 | 0 | 0 | 0 |
| **80** | 99 | 0 | 0 | 0 | 0 | 0 |
| **81** | 86 | 0 | 0 | 0 | 0 | 0 |
| **82** | 105 | 0 | 0 | 0 | 0 | 0 |
| **83** | 118 | 0 | 0 | 0 | 0 | 0 |
| **84** | 103 | 0 | 0 | 0 | 0 | 0 |
| **85** | 84 | 0 | 0 | 0 | 0 | 0 |
| **86** | 94 | 0 | 0 | 0 | 0 | 0 |
| **87** | 90 | 0 | 0 | 0 | 0 | 0 |
| **88** | 71 | 0 | 0 | 0 | 0 | 0 |
| **89** | 58 | 0 | 0 | 0 | 0 | 0 |
| **90** | 70 | 0 | 0 | 0 | 0 | 0 |
| **91** | 48 | 0 | 0 | 0 | 0 | 0 |
| **92** | 52 | 0 | 0 | 0 | 0 | 0 |
| **93** | 31 | 0 | 0 | 0 | 0 | 0 |
| **94** | 37 | 0 | 0 | 0 | 0 | 0 |
| **95** | 34 | 0 | 0 | 0 | 0 | 0 |
| **96** | 31 | 0 | 0 | 0 | 0 | 0 |
| **97** | 24 | 0 | 0 | 0 | 0 | 0 |
| **98** | 22 | 0 | 0 | 0 | 0 | 0 |
| **99** | 17 | 0 | 0 | 0 | 0 | 0 |
| **100** | 9 | 0 | 0 | 0 | 0 | 0 |
| **101** | 12 | 0 | 0 | 0 | 0 | 0 |
| **102** | 6 | 0 | 0 | 0 | 0 | 0 |
| **103** | 6 | 0 | 0 | 0 | 0 | 0 |
| **104** | 7 | 0 | 0 | 0 | 0 | 0 |
| **105** | 2 | 0 | 0 | 0 | 0 | 0 |
| **106** | 4 | 0 | 0 | 0 | 0 | 0 |
| **107** | 5 | 1 | 0 | 0 | 0 | 0 |
| **108** | 5 | 0 | 0 | 0 | 0 | 0 |
| **109** | 1 | 0 | 0 | 0 | 0 | 0 |
| **110** | 0 | 0 | 0 | 0 | 0 | 0 |
| **111** | 0 | 0 | 0 | 0 | 0 | 0 |
| **112** | 0 | 0 | 0 | 0 | 0 | 0 |
| **113** | 0 | 0 | 0 | 0 | 0 | 0 |
| **114** | 1 | 0 | 0 | 0 | 0 | 0 |
| **115** | 1 | 0 | 0 | 0 | 0 | 0 |
| **116** | 0 | 0 | 0 | 0 | 0 | 0 |
| **117** | 0 | 0 | 0 | 0 | 0 | 0 |
| **118** | 0 | 0 | 0 | 0 | 0 | 0 |
| **119** | 0 | 0 | 0 | 0 | 0 | 0 |
| **120** | 0 | 0 | 0 | 0 | 0 | 0 |

**Phase II Batch Experiments**

**Temp.** 20 °C

**Test 1**

| Time (min) | $[O_3]$ (mg/L) |
|---|---|
| 0 | 2.364 |
| 1.33 | 1.022 |
| 2.33 | 0.719 |
| 3.33 | 0.524 |
| 4.33 | 0.413 |
| 5.33 | 0.291 |
| 6.33 | 0.226 |
| 7.33 | 0.136 |
| 8.33 | 0.117 |

**Test 2**

| Time (min) | $[O_3]$ (mg/L) |
|---|---|
| 0 | 1.947 |
| 1.33 | 0.655 |
| 2.33 | 0.374 |
| 3.33 | 0.221 |
| 4.33 | 0.157 |
| 5.33 | 0.092 |

**Test 3**

| Time (min) | $[O_3]$ (mg/L) |
|---|---|
| 0 | 1.579 |
| 1.33 | 0.384 |
| 2.33 | 0.208 |
| 3.33 | 0.103 |
| 4.33 | 0.044 |

**Test 4**

| Time (min) | $[O_3]$ (mg/L) |
|---|---|
| 0 | 2.120 |
| 1.33 | 0.869 |
| 2.33 | 0.589 |
| 3.33 | 0.406 |
| 4.33 | 0.287 |
| 5.33 | 0.192 |
| 6.33 | 0.140 |
| 7.33 | 0.076 |
| 8.33 | 0.059 |

**Bromate Concentration (μg/L)**

| Test # | Conc. |
|---|---|
| **1** | 31.19 |
| **2** | 16.59 |
| **3** | 9.54 |
| **4** | 25.29 |

**Phase II Full-Scale Experiment**

**Dissolved Ozone Concentration (mg/L)**

| Chamber # | Port # | Test 1 | Test 2 | Test 3 | Test 4 | Test 5 | Test 6 |
|---|---|---|---|---|---|---|---|
| 1 | 1 | 0.190065 | 0.803456 | 0.319654 | 0.25054 | 0.682505 | 0.190065 |
| | 2 | 0.336933 | 0.907127 | 0.596112 | 0.613391 | 0.656587 | 0.259179 |
| | 3 | 0.552916 | 1.304536 | 0.777538 | 0.958963 | 1.010799 | 0.492441 |
| | 4 | 0.466523 | 1.330454 | 0.915767 | 0.941685 | 0.898488 | 0.535637 |
| | 5 | 0.388769 | 1.079914 | 0.604752 | 0.639309 | 0.613391 | 0.164147 |
| | b | 0.12095 | 0.717063 | 0.224622 | 0.440605 | 0.311015 | 0.233261 |
| 2 | 1 | 0.034557 | 0.267819 | 0.12959 | 0.302376 | 0.086393 | 0.846652 |
| | 2 | 0.017279 | 0.328294 | 0.112311 | 0.241901 | 0.112311 | 0.976242 |
| | 3 | 0 | 0.354212 | 0.025918 | 0.25054 | 0.146868 | 1.24406 |
| | 4 | 0 | 0.37149 | 0.138229 | 0.259179 | 0.155508 | 1.330454 |
| | 5 | 0.017279 | 0.37149 | 0.25054 | 0.233261 | 0.12959 | 0.950324 |
| | b | | | | | | |
| 3 | 1 | 0 | 0.138229 | 0 | 0.051836 | 0 | 0.38013 |
| | 2 | 0 | 0.112311 | 0 | 0.069114 | 0 | 0.336933 |
| | 3 | 0 | 0.112311 | 0 | 0.051836 | 0 | 0.38013 |
| | 4 | 0 | 0.164147 | 0 | 0.086393 | 0 | 0.50108 |
| | 5 | 0 | 0.112311 | 0 | 0.069114 | 0.017279 | 0.457883 |
| | b | 0 | 0.095032 | 0 | 0.034557 | 0 | 0.38013 |
| 4 | 1 | 0 | 0.086393 | 0 | 0.043197 | 0 | 0.172786 |
| | 2 | 0 | 0.034557 | 0 | 0.051836 | 0 | 0.12959 |
| | 3 | 0 | 0 | 0 | 0.043197 | 0 | 0.198704 |
| | 4 | 0 | 0 | 0 | 0.034557 | 0 | 0.198704 |
| | 5 | 0 | 0.043197 | 0 | 0.025918 | 0 | 0.224622 |
| | b | 0 | 0.017279 | 0 | 0.051836 | 0 | 0.12959 |
| 5 | 1 | 0 | 0 | 0 | 0 | 0 | 0.086393 |
| | 2 | 0 | 0 | 0 | 0 | 0 | 0.025918 |
| | 3 | 0 | 0 | 0 | 0 | 0 | 0.060475 |
| | 4 | 0 | 0 | 0 | 0 | 0 | 0.086393 |
| | 5 | 0 | 0 | 0 | 0 | 0 | 0.060475 |
| | b | 0 | 0 | 0 | 0 | 0 | 0.034557 |

**Phase II Full-Scale Experiment**

**Bromate Concentration (μg/L)**

| Chamber # | Port # | Test 1 | Test 2 | Test 3 | Test 4 | Test 5 | Test 6 |
|---|---|---|---|---|---|---|---|
| 1 | 1 | 4.080029 | 16.01147 | 11.14794 | 15.01981 | 12.21886 | 4.10659 |
| | 2 | 6.242266 | 23.42014 | 13.89179 | 15.92377 | 13.10635 | 4.378884 |
| | 3 | 7.283398 | 24.74273 | 14.40439 | 14.6948 | 14.45361 | |
| | 4 | 6.383271 | 23.21111 | 13.94191 | 16.06467 | 12.48393 | 4.814088 |
| | 5 | 6.867093 | 23.22383 | 16.41621 | 12.56013 | 10.64164 | 5.161049 |
| | b | 7.061024 | 30.87331 | 15.97592 | 13.59472 | 11.48951 | 17.23641 |
| 2 | 1 | 7.758805 | 32.30116 | 16.42322 | 12.69546 | 15.01288 | 17.7457 |
| | 2 | 7.138411 | 33.22233 | 16.46675 | 15.64648 | 14.00615 | 18.75536 |
| | 3 | 7.117725 | | 16.03936 | 16.86169 | 13.06076 | 16.8369 |
| | 4 | 7.08692 | 31.66921 | 16.71596 | 16.83666 | 11.27564 | 18.2965 |
| | 5 | 7.246828 | 32.49605 | 17.27597 | 15.80864 | 12.86607 | 16.1218 |
| | b | | | | | | |
| 3 | 1 | 6.991393 | 34.43087 | 16.86461 | 16.57239 | 13.45846 | 22.81843 |
| | 2 | 6.115629 | 36.44373 | 18.08475 | 16.16117 | 13.87429 | 23.51701 |
| | 3 | 7.285799 | 36.14702 | 17.22215 | 17.0228 | 13.79898 | 23.30464 |
| | 4 | 7.385673 | 34.1536 | 17.69861 | 17.09665 | 13.14342 | 24.34591 |
| | 5 | 7.145245 | 34.57334 | 17.50477 | 17.63113 | 13.57588 | 26.55452 |
| | b | 6.586355 | 37.5394 | 17.43711 | 17.03151 | 13.05443 | 26.81882 |
| 4 | 1 | 7.164696 | 36.17544 | 17.01461 | 16.66958 | 13.75122 | 26.64237 |
| | 2 | | | | 16.45639 | | 22.0128 |
| | 3 | 7.67717 | 36.01334 | 16.58572 | 16.54052 | 14.16103 | 27.18386 |
| | 4 | | | | 16.59904 | | 24.4456 |
| | 5 | 7.383503 | 34.85229 | 16.85449 | 16.9183 | 13.81816 | 23.4565 |
| | b | 7.379624 | 35.73364 | 17.91545 | 15.83912 | 13.72226 | 28.28193 |
| 5 | 1 | | 35.47638 | 17.10707 | 16.78122 | 13.06872 | 25.92634 |
| | 2 | | | | | | 25.46069 |
| | 3 | | 36.84477 | 16.63536 | 16.81414 | 13.0756 | 26.06908 |
| | 4 | 7.241329 | | | | | 24.92638 |
| | 5 | 7.735349 | 36.49496 | 16.71866 | 16.58894 | 13.79898 | 25.6141 |
| | b | 7.252184 | 36.31099 | 17.24779 | 17.44552 | 13.84693 | 25.63465 |

**Phase II Full-Scale Experiment**
**Test #** 1 **Chamber #** 1
**Pre-ozonated 0.94 μm YG-Microspheres**

| Intensity | Port #1 | Port #2 | Port #3 | Port #4 | Port #5 | Port #b |
|---|---|---|---|---|---|---|
| **1** | | 1019 | 1135 | 778 | 553 | 752 |
| **2** | | 2408 | 2596 | 1898 | 954 | 1843 |
| **3** | | 2193 | 2307 | 1597 | 719 | 1619 |
| **4** | | 928 | 1039 | 581 | 258 | 701 |
| **5** | | 224 | 287 | 150 | 93 | 204 |
| **6** | | 73 | 79 | 42 | 31 | 51 |
| **7** | | 24 | 26 | 15 | 18 | 27 |
| **8** | | 15 | 17 | 11 | 13 | 12 |
| **9** | | 5 | 13 | 10 | 9 | 9 |
| **10** | | 7 | 4 | 8 | 6 | 7 |
| **11** | | 3 | 1 | 9 | 15 | 3 |
| **12** | | 5 | 8 | 6 | 8 | 7 |
| **13** | | 2 | 4 | 6 | 9 | 2 |
| **14** | | 1 | 6 | 3 | 7 | 10 |
| **15** | | 6 | 9 | 10 | 11 | 9 |
| **16** | | 6 | 7 | 8 | 18 | 10 |
| **17** | | 3 | 11 | 7 | 25 | 9 |
| **18** | | 7 | 10 | 10 | 16 | 11 |
| **19** | | 11 | 10 | 5 | 17 | 17 |
| **20** | | 8 | 17 | 11 | 30 | 16 |
| **21** | | 9 | 17 | 12 | 31 | 30 |
| **22** | | 8 | 27 | 16 | 21 | 21 |
| **23** | | 14 | 30 | 22 | 32 | 38 |
| **24** | | 18 | 38 | 25 | 41 | 29 |
| **25** | | 17 | 33 | 36 | 37 | 54 |
| **26** | | 21 | 38 | 43 | 51 | 59 |
| **27** | | 24 | 59 | 44 | 42 | 68 |
| **28** | | 35 | 53 | 48 | 53 | 67 |
| **29** | | 35 | 73 | 59 | 71 | 54 |
| **30** | | 35 | 61 | 66 | 78 | 71 |
| **31** | | 42 | 80 | 70 | 71 | 96 |
| **32** | | 51 | 92 | 62 | 107 | 83 |
| **33** | | 56 | 94 | 83 | 80 | 96 |
| **34** | | 45 | 101 | 86 | 103 | 100 |
| **35** | | 67 | 100 | 82 | 80 | 108 |
| **36** | | 62 | 114 | 97 | 99 | 108 |
| **37** | | 43 | 104 | 76 | 82 | 94 |
| **38** | | 67 | 107 | 107 | 104 | 110 |
| **39** | | 50 | 109 | 100 | 111 | 110 |
| **40** | | 54 | 100 | 82 | 114 | 138 |
| **41** | | 67 | 113 | 101 | 111 | 103 |
| **42** | | 62 | 108 | 95 | 113 | 110 |
| **43** | | 81 | 107 | 99 | 88 | 115 |
| **44** | | 46 | 109 | 87 | 105 | 97 |
| **45** | | 66 | 104 | 94 | 94 | 92 |
| **46** | | 50 | 91 | 98 | 84 | 91 |
| **47** | | 57 | 87 | 104 | 123 | 80 |
| **48** | | 49 | 79 | 85 | 83 | 83 |
| **49** | | 46 | 104 | 77 | 79 | 90 |
| **50** | | 38 | 68 | 94 | 78 | 60 |
| **51** | | 49 | 61 | 73 | 64 | 59 |
| **52** | | 42 | 58 | 68 | 72 | 58 |
| **53** | | 43 | 58 | 73 | 63 | 48 |
| **54** | | 48 | 64 | 75 | 49 | 50 |
| **55** | | 47 | 45 | 56 | 58 | 39 |
| **56** | | 32 | 49 | 58 | 39 | 40 |
| **57** | | 26 | 44 | 42 | 28 | 51 |
| **58** | | 31 | 36 | 52 | 32 | 29 |
| **59** | | 22 | 34 | 48 | 31 | 27 |
| **60** | | 21 | 22 | 42 | 24 | 20 |
| **61** | | 20 | 25 | 40 | 22 | 21 |
| **62** | | 20 | 24 | 38 | 15 | 21 |
| **63** | | 18 | 11 | 30 | 25 | 16 |
| **64** | | 14 | 12 | 23 | 12 | 6 |
| **65** | | 13 | 10 | 16 | 9 | 11 |
| **66** | | 8 | 15 | 24 | 11 | 9 |
| **67** | | 10 | 9 | 23 | 8 | 6 |
| **68** | | 3 | 8 | 18 | 6 | 5 |
| **69** | | 7 | 4 | 7 | 3 | 9 |
| **70** | | 5 | 4 | 12 | 9 | 0 |
| **71** | | 7 | 1 | 14 | 5 | 3 |
| **72** | | 10 | 4 | 12 | 5 | 2 |
| **73** | | 3 | 1 | 6 | 1 | 2 |
| **74** | | 2 | 2 | 6 | 0 | 1 |
| **75** | | 4 | 0 | 4 | 0 | 0 |
| **76** | | 1 | 1 | 4 | 0 | 1 |
| **77** | | 2 | 1 | 1 | 1 | 0 |
| **78** | | 0 | 0 | 2 | 0 | 1 |
| **79** | | 1 | 0 | 2 | 0 | 1 |
| **80** | | 0 | 0 | 1 | 0 | 0 |
| **81** | | 0 | 2 | 0 | 1 | 1 |
| **82** | | 2 | 0 | 0 | 0 | 0 |
| **83** | | 0 | 0 | 2 | 0 | 1 |
| **84** | | 0 | 0 | 0 | 0 | 0 |
| **85** | | 0 | 0 | 0 | 0 | 0 |
| **86** | | 0 | 0 | 0 | 0 | 0 |
| **87** | | 0 | 0 | 0 | 0 | 0 |
| **88** | | 0 | 0 | 0 | 0 | 0 |
| **89** | | 0 | 0 | 0 | 0 | 0 |
| **90** | | 0 | 0 | 0 | 0 | 0 |
| **91** | | 1 | 0 | 0 | 0 | 0 |
| **92** | | 0 | 0 | 0 | 0 | 0 |
| **93** | | 0 | 0 | 0 | 0 | 0 |
| **94** | | 0 | 0 | 0 | 0 | 0 |
| **95** | | 0 | 0 | 0 | 0 | 0 |
| **96** | | 0 | 0 | 0 | 0 | 0 |
| **97** | | 0 | 0 | 0 | 0 | 0 |
| **98** | | 0 | 0 | 0 | 0 | 0 |
| **99** | | 0 | 0 | 0 | 0 | 0 |
| **100** | | 0 | 0 | 0 | 0 | 0 |

**Phase II Full-Scale Experiment**
**Test #** 1 **Chamber #** 2
**Pre-ozonated 0.94 μm YG-Microspheres**

| Intensity | Port #1 | Port #2 | Port #3 | Port #4 | Port #5 | Port #b |
|---|---|---|---|---|---|---|
| 1 | 723 | 478 | 759 | 911 | 909 | |
| 2 | 1398 | 884 | 1373 | 1863 | 1897 | |
| 3 | 1035 | 622 | 1061 | 1609 | 1496 | |
| 4 | 393 | 228 | 427 | 634 | 654 | |
| 5 | 134 | 92 | 143 | 226 | 184 | |
| 6 | 44 | 26 | 39 | 88 | 79 | |
| 7 | 22 | 18 | 34 | 39 | 43 | |
| 8 | 20 | 7 | 25 | 21 | 26 | |
| 9 | 17 | 7 | 18 | 14 | 18 | |
| 10 | 6 | 7 | 6 | 15 | 14 | |
| 11 | 7 | 8 | 7 | 14 | 13 | |
| 12 | 4 | 5 | 6 | 13 | 14 | |
| 13 | 6 | 5 | 6 | 5 | 8 | |
| 14 | 7 | 6 | 10 | 7 | 12 | |
| 15 | 8 | 7 | 7 | 10 | 10 | |
| 16 | 10 | 15 | 10 | 11 | 16 | |
| 17 | 13 | 16 | 12 | 20 | 11 | |
| 18 | 12 | 7 | 15 | 12 | 11 | |
| 19 | 9 | 14 | 12 | 22 | 21 | |
| 20 | 22 | 25 | 16 | 19 | 24 | |
| 21 | 21 | 24 | 25 | 26 | 25 | |
| 22 | 19 | 38 | 21 | 31 | 36 | |
| 23 | 20 | 38 | 38 | 38 | 30 | |
| 24 | 33 | 30 | 37 | 47 | 57 | |
| 25 | 53 | 49 | 53 | 43 | 54 | |
| 26 | 45 | 49 | 44 | 66 | 66 | |
| 27 | 60 | 55 | 80 | 62 | 72 | |
| 28 | 65 | 69 | 74 | 88 | 73 | |
| 29 | 75 | 73 | 81 | 66 | 94 | |
| 30 | 75 | 78 | 101 | 81 | 75 | |
| 31 | 73 | 87 | 74 | 100 | 97 | |
| 32 | 87 | 102 | 109 | 102 | 83 | |
| 33 | 83 | 91 | 107 | 97 | 110 | |
| 34 | 102 | 86 | 112 | 117 | 105 | |
| 35 | 113 | 93 | 120 | 114 | 103 | |
| 36 | 119 | 111 | 103 | 110 | 119 | |
| 37 | 115 | 114 | 106 | 119 | 113 | |
| 38 | 107 | 118 | 120 | 127 | 128 | |
| 39 | 108 | 134 | 102 | 126 | 107 | |
| 40 | 104 | 99 | 133 | 106 | 105 | |
| 41 | 113 | 108 | 101 | 123 | 107 | |
| 42 | 109 | 118 | 86 | 104 | 98 | |
| 43 | 99 | 99 | 89 | 90 | 107 | |
| 44 | 108 | 97 | 102 | 100 | 96 | |
| 45 | 99 | 96 | 100 | 80 | 90 | |
| 46 | 88 | 107 | 76 | 100 | 89 | |
| 47 | 90 | 82 | 101 | 79 | 84 | |
| 48 | 61 | 72 | 86 | 81 | 85 | |
| 49 | 70 | 83 | 74 | 84 | 86 | |
| 50 | 88 | 72 | 74 | 79 | 65 | |

| Intensity | Port #1 | Port #2 | Port #3 | Port #4 | Port #5 | Port #b |
|---|---|---|---|---|---|---|
| 51 | 58 | 69 | 62 | 71 | 57 | |
| 52 | 71 | 54 | 55 | 53 | 64 | |
| 53 | 60 | 64 | 54 | 56 | 56 | |
| 54 | 57 | 52 | 42 | 39 | 44 | |
| 55 | 38 | 46 | 42 | 48 | 39 | |
| 56 | 40 | 30 | 33 | 32 | 36 | |
| 57 | 42 | 36 | 39 | 33 | 26 | |
| 58 | 31 | 28 | 39 | 28 | 26 | |
| 59 | 29 | 29 | 32 | 30 | 30 | |
| 60 | 18 | 26 | 17 | 20 | 22 | |
| 61 | 24 | 23 | 15 | 23 | 20 | |
| 62 | 17 | 20 | 12 | 12 | 17 | |
| 63 | 23 | 9 | 16 | 9 | 13 | |
| 64 | 13 | 9 | 6 | 14 | 17 | |
| 65 | 8 | 8 | 14 | 8 | 9 | |
| 66 | 5 | 10 | 12 | 6 | 5 | |
| 67 | 9 | 9 | 5 | 8 | 5 | |
| 68 | 9 | 6 | 6 | 5 | 3 | |
| 69 | 7 | 1 | 7 | 7 | 2 | |
| 70 | 4 | 1 | 3 | 1 | 1 | |
| 71 | 1 | 3 | 1 | 1 | 1 | |
| 72 | 1 | 1 | 3 | 2 | 2 | |
| 73 | 3 | 4 | 2 | 2 | 0 | |
| 74 | 4 | 3 | 0 | 1 | 1 | |
| 75 | 2 | 2 | 1 | 1 | 1 | |
| 76 | 1 | 2 | 1 | 3 | 3 | |
| 77 | 0 | 0 | 1 | 0 | 2 | |
| 78 | 1 | 0 | 2 | 1 | 0 | |
| 79 | 0 | 0 | 0 | 2 | 0 | |
| 80 | 0 | 0 | 0 | 1 | 0 | |
| 81 | 0 | 0 | 1 | 0 | 0 | |
| 82 | 0 | 0 | 0 | 0 | 0 | |
| 83 | 0 | 0 | 0 | 0 | 0 | |
| 84 | 0 | 0 | 0 | 0 | 1 | |
| 85 | 0 | 0 | 1 | 1 | 0 | |
| 86 | 0 | 0 | 0 | 0 | 0 | |
| 87 | 0 | 0 | 0 | 0 | 1 | |
| 88 | 0 | 0 | 0 | 0 | 0 | |
| 89 | 0 | 0 | 0 | 0 | 0 | |
| 90 | 1 | 0 | 0 | 0 | 0 | |
| 91 | 0 | 0 | 0 | 0 | 0 | |
| 92 | 0 | 0 | 0 | 0 | 0 | |
| 93 | 0 | 0 | 0 | 0 | 0 | |
| 94 | 0 | 0 | 0 | 0 | 0 | |
| 95 | 0 | 0 | 0 | 0 | 0 | |
| 96 | 0 | 0 | 0 | 0 | 0 | |
| 97 | 0 | 0 | 0 | 0 | 0 | |
| 98 | 0 | 0 | 0 | 0 | 0 | |
| 99 | 0 | 0 | 0 | 0 | 0 | |
| 100 | 0 | 0 | 0 | 0 | 0 | |

**Phase II Full-Scale Experiment**
**Test #** 1 **Chamber #** 3
**Pre-ozonated 0.94 $\mu$m YG-Microspheres**

| Intensity | Port #1 | Port #2 | Port #3 | Port #4 | Port #5 | Port #b |
|---|---|---|---|---|---|---|
| **1** | 1058 | 1508 | 797 | | 872 | 1371 |
| **2** | 2372 | 3111 | 1427 | | 1766 | 2904 |
| **3** | 1940 | 2586 | 1109 | | 1539 | 2533 |
| **4** | 774 | 1059 | 420 | | 602 | 1038 |
| **5** | 260 | 309 | 145 | | 207 | 298 |
| **6** | 85 | 121 | 56 | | 63 | 101 |
| **7** | 37 | 59 | 34 | | 35 | 44 |
| **8** | 20 | 32 | 14 | | 10 | 34 |
| **9** | 16 | 16 | 12 | | 12 | 18 |
| **10** | 9 | 14 | 5 | | 12 | 10 |
| **11** | 9 | 17 | 6 | | 9 | 8 |
| **12** | 13 | 13 | 7 | | 5 | 9 |
| **13** | 9 | 13 | 9 | | 7 | 10 |
| **14** | 3 | 18 | 5 | | 7 | 6 |
| **15** | 7 | 12 | 8 | | 7 | 11 |
| **16** | 4 | 11 | 10 | | 11 | 10 |
| **17** | 13 | 14 | 13 | | 11 | 12 |
| **18** | 10 | 16 | 13 | | 19 | 10 |
| **19** | 16 | 18 | 15 | | 19 | 17 |
| **20** | 15 | 26 | 21 | | 23 | 20 |
| **21** | 32 | 19 | 35 | | 32 | 31 |
| **22** | 37 | 40 | 35 | | 26 | 30 |
| **23** | 35 | 42 | 37 | | 47 | 36 |
| **24** | 28 | 39 | 50 | | 49 | 48 |
| **25** | 39 | 61 | 51 | | 40 | 66 |
| **26** | 63 | 44 | 61 | | 61 | 75 |
| **27** | 73 | 64 | 83 | | 44 | 75 |
| **28** | 88 | 65 | 65 | | 73 | 83 |
| **29** | 74 | 89 | 78 | | 90 | 76 |
| **30** | 90 | 79 | 88 | | 99 | 106 |
| **31** | 80 | 96 | 94 | | 90 | 112 |
| **32** | 82 | 97 | 86 | | 109 | 78 |
| **33** | 97 | 105 | 105 | | 125 | 116 |
| **34** | 102 | 96 | 98 | | 108 | 100 |
| **35** | 106 | 113 | 107 | | 131 | 119 |
| **36** | 100 | 110 | 117 | | 115 | 112 |
| **37** | 126 | 99 | 112 | | 123 | 96 |
| **38** | 111 | 117 | 127 | | 108 | 106 |
| **39** | 123 | 114 | 104 | | 100 | 124 |
| **40** | 88 | 92 | 109 | | 91 | 112 |
| **41** | 96 | 117 | 102 | | 110 | 114 |
| **42** | 112 | 108 | 97 | | 101 | 86 |
| **43** | 91 | 108 | 85 | | 113 | 102 |
| **44** | 85 | 106 | 86 | | 94 | 99 |
| **45** | 95 | 96 | 93 | | 90 | 97 |
| **46** | 88 | 79 | 98 | | 81 | 91 |
| **47** | 86 | 81 | 74 | | 83 | 101 |
| **48** | 88 | 63 | 65 | | 77 | 84 |
| **49** | 75 | 83 | 71 | | 70 | 74 |
| **50** | 70 | 68 | 77 | | 71 | 78 |

| Intensity | Port #1 | Port #2 | Port #3 | Port #4 | Port #5 | Port #b |
|---|---|---|---|---|---|---|
| **51** | 68 | 72 | 73 | | 55 | 55 |
| **52** | 53 | 55 | 52 | | 56 | 64 |
| **53** | 57 | 45 | 49 | | 46 | 45 |
| **54** | 45 | 38 | 43 | | 52 | 52 |
| **55** | 38 | 34 | 37 | | 43 | 37 |
| **56** | 34 | 34 | 48 | | 39 | 38 |
| **57** | 34 | 29 | 20 | | 34 | 43 |
| **58** | 27 | 27 | 25 | | 29 | 27 |
| **59** | 31 | 21 | 26 | | 17 | 27 |
| **60** | 24 | 22 | 25 | | 18 | 24 |
| **61** | 21 | 16 | 18 | | 19 | 12 |
| **62** | 25 | 13 | 12 | | 13 | 21 |
| **63** | 16 | 14 | 15 | | 15 | 23 |
| **64** | 10 | 8 | 14 | | 8 | 18 |
| **65** | 8 | 13 | 12 | | 11 | 12 |
| **66** | 10 | 8 | 8 | | 13 | 6 |
| **67** | 6 | 5 | 7 | | 8 | 12 |
| **68** | 9 | 5 | 7 | | 6 | 5 |
| **69** | 2 | 3 | 2 | | 2 | 3 |
| **70** | 5 | 5 | 2 | | 2 | 2 |
| **71** | 5 | 2 | 0 | | 1 | 2 |
| **72** | 0 | 5 | 3 | | 0 | 0 |
| **73** | 4 | 3 | 1 | | 1 | 3 |
| **74** | 2 | 1 | 0 | | 1 | 3 |
| **75** | 1 | 1 | 0 | | 0 | 0 |
| **76** | 3 | 0 | 1 | | 0 | 1 |
| **77** | 1 | 0 | 0 | | 1 | 0 |
| **78** | 2 | 2 | 0 | | 0 | 1 |
| **79** | 0 | 0 | 0 | | 0 | 0 |
| **80** | 0 | 0 | 1 | | 0 | 0 |
| **81** | 0 | 0 | 0 | | 0 | 0 |
| **82** | 0 | 0 | 0 | | 0 | 0 |
| **83** | 0 | 0 | 0 | | 0 | 1 |
| **84** | 0 | 0 | 0 | | 0 | 0 |
| **85** | 0 | 0 | 0 | | 0 | 0 |
| **86** | 0 | 0 | 0 | | 0 | 0 |
| **87** | 0 | 0 | 0 | | 0 | 0 |
| **88** | 0 | 0 | 0 | | 0 | 0 |
| **89** | 0 | 0 | 0 | | 0 | 0 |
| **90** | 0 | 0 | 0 | | 0 | 0 |
| **91** | 0 | 0 | 0 | | 1 | 0 |
| **92** | 0 | 0 | 0 | | 0 | 0 |
| **93** | 0 | 0 | 0 | | 0 | 0 |
| **94** | 0 | 0 | 0 | | 0 | 0 |
| **95** | 0 | 0 | 0 | | 0 | 0 |
| **96** | 0 | 0 | 0 | | 0 | 0 |
| **97** | 0 | 0 | 0 | | 0 | 0 |
| **98** | 0 | 0 | 0 | | 0 | 0 |
| **99** | 0 | 0 | 0 | | 0 | 0 |
| **100** | 0 | 0 | 0 | | 0 | 0 |

**Phase II Full-Scale Experiment**
**Test #** 1 **Chamber #** 4
**Pre-ozonated 0.94 μm YG-Microspheres**

| Intensity | Port #1 | Port #2 | Port #3 | Port #4 | Port #5 | Port #b |
|---|---|---|---|---|---|---|
| 1 | 1091 | 804 | 673 | 409 | 745 | 746 |
| 2 | 2163 | 1513 | 1400 | 787 | 1511 | 1430 |
| 3 | 1700 | 1144 | 1105 | 525 | 1224 | 1140 |
| 4 | 639 | 426 | 426 | 204 | 460 | 451 |
| 5 | 227 | 142 | 114 | 69 | 137 | 144 |
| 6 | 67 | 53 | 45 | 27 | 68 | 69 |
| 7 | 35 | 21 | 15 | 17 | 29 | 24 |
| 8 | 22 | 13 | 14 | 6 | 14 | 13 |
| 9 | 10 | 11 | 10 | 10 | 6 | 11 |
| 10 | 10 | 7 | 4 | 14 | 7 | 3 |
| 11 | 5 | 6 | 9 | 7 | 11 | 5 |
| 12 | 3 | 6 | 8 | 9 | 4 | 4 |
| 13 | 8 | 6 | 9 | 10 | 5 | 8 |
| 14 | 11 | 7 | 2 | 7 | 10 | 7 |
| 15 | 10 | 9 | 5 | 8 | 3 | 11 |
| 16 | 13 | 15 | 12 | 16 | 11 | 12 |
| 17 | 12 | 7 | 14 | 13 | 19 | 11 |
| 18 | 17 | 17 | 11 | 10 | 18 | 17 |
| 19 | 15 | 29 | 15 | 19 | 10 | 17 |
| 20 | 28 | 23 | 22 | 22 | 12 | 22 |
| 21 | 33 | 27 | 26 | 16 | 26 | 18 |
| 22 | 35 | 40 | 33 | 41 | 35 | 37 |
| 23 | 42 | 42 | 55 | 40 | 25 | 51 |
| 24 | 48 | 58 | 49 | 49 | 42 | 55 |
| 25 | 48 | 61 | 58 | 53 | 54 | 58 |
| 26 | 60 | 67 | 54 | 67 | 44 | 71 |
| 27 | 79 | 82 | 71 | 74 | 68 | 65 |
| 28 | 72 | 86 | 71 | 59 | 66 | 95 |
| 29 | 85 | 85 | 88 | 93 | 61 | 82 |
| 30 | 74 | 100 | 65 | 80 | 89 | 89 |
| 31 | 95 | 113 | 96 | 91 | 86 | 88 |
| 32 | 117 | 118 | 102 | 111 | 96 | 91 |
| 33 | 108 | 103 | 97 | 107 | 108 | 104 |
| 34 | 100 | 129 | 95 | 127 | 120 | 123 |
| 35 | 123 | 114 | 114 | 111 | 109 | 120 |
| 36 | 110 | 110 | 111 | 109 | 119 | 103 |
| 37 | 115 | 141 | 129 | 113 | 126 | 120 |
| 38 | 105 | 129 | 89 | 116 | 125 | 99 |
| 39 | 125 | 118 | 110 | 101 | 115 | 115 |
| 40 | 107 | 124 | 95 | 103 | 102 | 104 |
| 41 | 120 | 134 | 124 | 117 | 118 | 112 |
| 42 | 108 | 110 | 105 | 118 | 103 | 122 |
| 43 | 87 | 106 | 100 | 89 | 82 | 113 |
| 44 | 96 | 101 | 86 | 100 | 81 | 101 |
| 45 | 97 | 104 | 74 | 88 | 105 | 87 |
| 46 | 102 | 113 | 79 | 110 | 80 | 92 |
| 47 | 88 | 89 | 85 | 75 | 75 | 90 |
| 48 | 104 | 85 | 88 | 84 | 65 | 96 |
| 49 | 74 | 83 | 57 | 74 | 75 | 68 |
| 50 | 90 | 81 | 76 | 57 | 69 | 80 |

| Intensity | Port #1 | Port #2 | Port #3 | Port #4 | Port #5 | Port #b |
|---|---|---|---|---|---|---|
| 51 | 73 | 82 | 69 | 61 | 52 | 66 |
| 52 | 70 | 64 | 50 | 52 | 42 | 66 |
| 53 | 49 | 58 | 57 | 55 | 56 | 54 |
| 54 | 50 | 41 | 32 | 51 | 53 | 51 |
| 55 | 45 | 66 | 38 | 42 | 47 | 53 |
| 56 | 34 | 37 | 32 | 43 | 56 | 35 |
| 57 | 37 | 45 | 29 | 29 | 34 | 40 |
| 58 | 33 | 32 | 28 | 35 | 39 | 26 |
| 59 | 40 | 34 | 18 | 26 | 31 | 29 |
| 60 | 32 | 28 | 19 | 25 | 24 | 24 |
| 61 | 26 | 23 | 22 | 16 | 26 | 17 |
| 62 | 17 | 30 | 20 | 13 | 14 | 16 |
| 63 | 20 | 15 | 16 | 17 | 17 | 13 |
| 64 | 12 | 11 | 9 | 11 | 12 | 11 |
| 65 | 11 | 13 | 14 | 10 | 7 | 12 |
| 66 | 10 | 7 | 9 | 4 | 10 | 10 |
| 67 | 3 | 8 | 3 | 9 | 5 | 11 |
| 68 | 7 | 14 | 6 | 7 | 3 | 1 |
| 69 | 9 | 7 | 3 | 6 | 7 | 4 |
| 70 | 4 | 6 | 4 | 5 | 2 | 2 |
| 71 | 3 | 6 | 5 | 4 | 2 | 3 |
| 72 | 5 | 1 | 1 | 1 | 1 | 3 |
| 73 | 1 | 2 | 3 | 1 | 0 | 0 |
| 74 | 1 | 2 | 2 | 0 | 2 | 0 |
| 75 | 1 | 1 | 0 | 2 | 0 | 1 |
| 76 | 0 | 2 | 1 | 0 | 0 | 0 |
| 77 | 1 | 2 | 0 | 0 | 0 | 0 |
| 78 | 2 | 2 | 0 | 0 | 0 | 0 |
| 79 | 1 | 0 | 0 | 0 | 0 | 0 |
| 80 | 0 | 0 | 0 | 0 | 0 | 1 |
| 81 | 1 | 0 | 0 | 0 | 0 | 0 |
| 82 | 0 | 0 | 0 | 0 | 0 | 1 |
| 83 | 0 | 0 | 0 | 0 | 0 | 0 |
| 84 | 0 | 0 | 0 | 1 | 0 | 0 |
| 85 | 0 | 0 | 0 | 0 | 0 | 0 |
| 86 | 0 | 0 | 0 | 0 | 0 | 0 |
| 87 | 0 | 0 | 0 | 0 | 0 | 0 |
| 88 | 0 | 0 | 0 | 0 | 0 | 0 |
| 89 | 0 | 0 | 0 | 0 | 0 | 0 |
| 90 | 0 | 0 | 0 | 0 | 0 | 0 |
| 91 | 0 | 0 | 0 | 0 | 0 | 0 |
| 92 | 0 | 0 | 0 | 0 | 0 | 0 |
| 93 | 0 | 0 | 0 | 0 | 0 | 0 |
| 94 | 0 | 0 | 0 | 0 | 0 | 1 |
| 95 | 0 | 0 | 0 | 0 | 0 | 0 |
| 96 | 0 | 0 | 0 | 0 | 0 | 0 |
| 97 | 0 | 0 | 0 | 0 | 0 | 0 |
| 98 | 0 | 0 | 0 | 0 | 0 | 0 |
| 99 | 0 | 0 | 0 | 0 | 0 | 0 |
| 100 | 0 | 0 | 0 | 0 | 0 | 0 |

**Phase II Full-Scale Experiment**
**Test #** 1 **Chamber #** 5
**Pre-ozonated 0.94 μm YG-Microspheres**

| Intensity | Port #1 | Port #2 | Port #3 | Port #4 | Port #5 | Port #b |
|---|---|---|---|---|---|---|
| **1** | | 578 | | 694 | 684 | 733 |
| **2** | | 1090 | | 1260 | 1216 | 1475 |
| **3** | | 746 | | 923 | 887 | 1146 |
| **4** | | 275 | | 347 | 325 | 474 |
| **5** | | 115 | | 110 | 115 | 163 |
| **6** | | 60 | | 42 | 48 | 52 |
| **7** | | 45 | | 29 | 23 | 27 |
| **8** | | 16 | | 11 | 16 | 16 |
| **9** | | 22 | | 7 | 10 | 14 |
| **10** | | 20 | | 9 | 10 | 10 |
| **11** | | 15 | | 4 | 7 | 6 |
| **12** | | 12 | | 6 | 4 | 6 |
| **13** | | 8 | | 11 | 9 | 7 |
| **14** | | 13 | | 7 | 8 | 9 |
| **15** | | 12 | | 4 | 4 | 7 |
| **16** | | 3 | | 13 | 4 | 11 |
| **17** | | 11 | | 8 | 15 | 16 |
| **18** | | 13 | | 12 | 13 | 18 |
| **19** | | 28 | | 19 | 21 | 27 |
| **20** | | 40 | | 28 | 28 | 25 |
| **21** | | 41 | | 26 | 22 | 32 |
| **22** | | 29 | | 41 | 29 | 25 |
| **23** | | 34 | | 55 | 45 | 39 |
| **24** | | 60 | | 45 | 50 | 43 |
| **25** | | 69 | | 54 | 71 | 57 |
| **26** | | 65 | | 59 | 62 | 56 |
| **27** | | 85 | | 68 | 68 | 79 |
| **28** | | 77 | | 71 | 68 | 72 |
| **29** | | 88 | | 74 | 82 | 86 |
| **30** | | 105 | | 87 | 102 | 100 |
| **31** | | 112 | | 79 | 119 | 90 |
| **32** | | 109 | | 94 | 107 | 118 |
| **33** | | 100 | | 111 | 106 | 100 |
| **34** | | 95 | | 118 | 110 | 108 |
| **35** | | 107 | | 129 | 118 | 119 |
| **36** | | 103 | | 106 | 125 | 96 |
| **37** | | 96 | | 123 | 97 | 106 |
| **38** | | 126 | | 113 | 113 | 100 |
| **39** | | 98 | | 98 | 129 | 115 |
| **40** | | 99 | | 120 | 119 | 112 |
| **41** | | 122 | | 127 | 106 | 111 |
| **42** | | 106 | | 113 | 107 | 110 |
| **43** | | 109 | | 95 | 95 | 103 |
| **44** | | 98 | | 94 | 97 | 89 |
| **45** | | 88 | | 84 | 88 | 73 |
| **46** | | 94 | | 73 | 84 | 81 |
| **47** | | 86 | | 80 | 92 | 72 |
| **48** | | 77 | | 70 | 89 | 87 |
| **49** | | 75 | | 80 | 73 | 78 |
| **50** | | 69 | | 49 | 65 | 61 |

| Intensity | Port #1 | Port #2 | Port #3 | Port #4 | Port #5 | Port #b |
|---|---|---|---|---|---|---|
| **51** | | 58 | | 61 | 75 | 49 |
| **52** | | 62 | | 60 | 50 | 61 |
| **53** | | 55 | | 47 | 44 | 62 |
| **54** | | 46 | | 33 | 33 | 51 |
| **55** | | 35 | | 30 | 47 | 36 |
| **56** | | 42 | | 36 | 44 | 32 |
| **57** | | 33 | | 31 | 25 | 29 |
| **58** | | 23 | | 26 | 35 | 27 |
| **59** | | 23 | | 24 | 19 | 31 |
| **60** | | 23 | | 20 | 17 | 24 |
| **61** | | 14 | | 24 | 21 | 15 |
| **62** | | 14 | | 14 | 19 | 7 |
| **63** | | 15 | | 10 | 6 | 15 |
| **64** | | 7 | | 9 | 15 | 9 |
| **65** | | 12 | | 7 | 9 | 5 |
| **66** | | 10 | | 4 | 3 | 10 |
| **67** | | 7 | | 5 | 6 | 6 |
| **68** | | 8 | | 2 | 9 | 2 |
| **69** | | 2 | | 6 | 4 | 1 |
| **70** | | 5 | | 3 | 4 | 1 |
| **71** | | 2 | | 0 | 0 | 1 |
| **72** | | 2 | | 4 | 3 | 0 |
| **73** | | 2 | | 1 | 1 | 3 |
| **74** | | 1 | | 1 | 3 | 2 |
| **75** | | 1 | | 2 | 3 | 1 |
| **76** | | 2 | | 1 | 4 | 0 |
| **77** | | 0 | | 0 | 0 | 0 |
| **78** | | 0 | | 0 | 0 | 0 |
| **79** | | 2 | | 0 | 2 | 0 |
| **80** | | 0 | | 0 | 0 | 0 |
| **81** | | 1 | | 0 | 0 | 0 |
| **82** | | 0 | | 0 | 0 | 0 |
| **83** | | 0 | | 0 | 1 | 0 |
| **84** | | 0 | | 0 | 0 | 0 |
| **85** | | 0 | | 0 | 0 | 0 |
| **86** | | 0 | | 0 | 0 | 0 |
| **87** | | 0 | | 0 | 0 | 0 |
| **88** | | 0 | | 0 | 0 | 0 |
| **89** | | 0 | | 0 | 0 | 0 |
| **90** | | 0 | | 0 | 0 | 0 |
| **91** | | 0 | | 0 | 0 | 0 |
| **92** | | 0 | | 0 | 0 | 0 |
| **93** | | 0 | | 0 | 0 | 0 |
| **94** | | 0 | | 0 | 0 | 0 |
| **95** | | 0 | | 0 | 0 | 0 |
| **96** | | 0 | | 0 | 0 | 0 |
| **97** | | 0 | | 0 | 1 | 0 |
| **98** | | 0 | | 0 | 0 | 0 |
| **99** | | 0 | | 0 | 0 | 0 |
| **100** | | 0 | | 0 | 0 | 0 |

**Phase II Full-Scale Experiment**
**Test #** 2 **Chamber #** 1
**Pre-ozonated 0.94 μm YG-Microspheres**

| Intensity | Port #1 | Port #2 | Port #3 | Port #4 | Port #5 | Port #b |
|---|---|---|---|---|---|---|
| 1 | 12 | 8 | 4 | 9 | 4 | 3 |
| 2 | 53 | 49 | 40 | 70 | 48 | 29 |
| 3 | 102 | 108 | 56 | 113 | 112 | 63 |
| 4 | 106 | 90 | 52 | 110 | 81 | 72 |
| 5 | 67 | 58 | 28 | 53 | 53 | 30 |
| 6 | 33 | 32 | 23 | 36 | 35 | 18 |
| 7 | 17 | 23 | 15 | 24 | 29 | 23 |
| 8 | 23 | 22 | 19 | 14 | 20 | 13 |
| 9 | 14 | 13 | 14 | 11 | 24 | 26 |
| 10 | 14 | 19 | 23 | 13 | 20 | 20 |
| 11 | 20 | 21 | 22 | 25 | 22 | 25 |
| 12 | 18 | 21 | 28 | 22 | 27 | 33 |
| 13 | 18 | 38 | 21 | 28 | 21 | 37 |
| 14 | 20 | 33 | 45 | 36 | 39 | 42 |
| 15 | 31 | 34 | 49 | 37 | 43 | 49 |
| 16 | 27 | 40 | 37 | 39 | 47 | 66 |
| 17 | 31 | 69 | 60 | 55 | 46 | 55 |
| 18 | 30 | 73 | 51 | 52 | 73 | 81 |
| 19 | 45 | 74 | 71 | 66 | 74 | 91 |
| 20 | 55 | 76 | 70 | 61 | 85 | 97 |
| 21 | 52 | 84 | 90 | 76 | 83 | 108 |
| 22 | 44 | 73 | 94 | 82 | 90 | 128 |
| 23 | 60 | 88 | 89 | 74 | 96 | 97 |
| 24 | 61 | 108 | 109 | 79 | 97 | 122 |
| 25 | 75 | 93 | 102 | 96 | 96 | 145 |
| 26 | 65 | 98 | 103 | 94 | 115 | 128 |
| 27 | 75 | 114 | 111 | 134 | 127 | 138 |
| 28 | 76 | 130 | 120 | 123 | 123 | 134 |
| 29 | 76 | 146 | 114 | 85 | 108 | 153 |
| 30 | 84 | 123 | 112 | 115 | 107 | 128 |
| 31 | 79 | 106 | 133 | 118 | 121 | 139 |
| 32 | 77 | 116 | 113 | 93 | 119 | 130 |
| 33 | 114 | 103 | 109 | 117 | 115 | 118 |
| 34 | 95 | 115 | 129 | 100 | 90 | 102 |
| 35 | 109 | 124 | 114 | 103 | 117 | 97 |
| 36 | 96 | 80 | 83 | 101 | 87 | 75 |
| 37 | 89 | 93 | 106 | 94 | 105 | 94 |
| 38 | 88 | 86 | 81 | 89 | 101 | 92 |
| 39 | 89 | 96 | 90 | 93 | 88 | 84 |
| 40 | 99 | 95 | 80 | 88 | 91 | 79 |
| 41 | 88 | 80 | 85 | 86 | 83 | 55 |
| 42 | 77 | 78 | 85 | 69 | 68 | 57 |
| 43 | 71 | 65 | 67 | 82 | 70 | 45 |
| 44 | 77 | 66 | 66 | 65 | 73 | 41 |
| 45 | 60 | 49 | 59 | 66 | 58 | 36 |
| 46 | 74 | 46 | 56 | 49 | 41 | 33 |
| 47 | 68 | 47 | 41 | 61 | 54 | 14 |
| 48 | 70 | 46 | 45 | 52 | 47 | 28 |
| 49 | 66 | 39 | 36 | 46 | 33 | 15 |
| 50 | 64 | 39 | 35 | 43 | 34 | 18 |

| Intensity | Port #1 | Port #2 | Port #3 | Port #4 | Port #5 | Port #b |
|---|---|---|---|---|---|---|
| 51 | 56 | 33 | 29 | 35 | 34 | 13 |
| 52 | 45 | 27 | 24 | 38 | 32 | 11 |
| 53 | 54 | 35 | 12 | 37 | 25 | 9 |
| 54 | 46 | 21 | 20 | 36 | 22 | 10 |
| 55 | 46 | 24 | 10 | 26 | 14 | 8 |
| 56 | 45 | 20 | 12 | 24 | 12 | 4 |
| 57 | 41 | 19 | 10 | 17 | 8 | 6 |
| 58 | 37 | 9 | 8 | 16 | 8 | 1 |
| 59 | 47 | 9 | 9 | 25 | 9 | 0 |
| 60 | 28 | 7 | 7 | 16 | 7 | 2 |
| 61 | 23 | 11 | 7 | 5 | 6 | 1 |
| 62 | 29 | 2 | 4 | 6 | 5 | 0 |
| 63 | 24 | 2 | 3 | 6 | 4 | 2 |
| 64 | 17 | 4 | 2 | 2 | 3 | 0 |
| 65 | 22 | 1 | 1 | 2 | 2 | 1 |
| 66 | 16 | 1 | 0 | 2 | 2 | 1 |
| 67 | 17 | 2 | 0 | 0 | 2 | 3 |
| 68 | 11 | 1 | 0 | 2 | 2 | 1 |
| 69 | 5 | 0 | 0 | 1 | 0 | 0 |
| 70 | 3 | 0 | 0 | 0 | 0 | 0 |
| 71 | 3 | 0 | 0 | 0 | 1 | 0 |
| 72 | 5 | 0 | 0 | 1 | 0 | 0 |
| 73 | 2 | 1 | 0 | 0 | 1 | 0 |
| 74 | 2 | 0 | 0 | 3 | 1 | 0 |
| 75 | 4 | 0 | 0 | 0 | 0 | 0 |
| 76 | 3 | 0 | 0 | 0 | 0 | 0 |
| 77 | 1 | 0 | 0 | 1 | 0 | 1 |
| 78 | 0 | 0 | 0 | 0 | 0 | 0 |
| 79 | 1 | 0 | 0 | 0 | 0 | 0 |
| 80 | 0 | 0 | 0 | 0 | 0 | 0 |
| 81 | 0 | 0 | 0 | 0 | 0 | 1 |
| 82 | 0 | 0 | 0 | 0 | 0 | 0 |
| 83 | 0 | 0 | 0 | 0 | 0 | 0 |
| 84 | 0 | 0 | 0 | 0 | 0 | 0 |
| 85 | 0 | 0 | 0 | 0 | 0 | 0 |
| 86 | 1 | 0 | 0 | 0 | 0 | 0 |
| 87 | 0 | 0 | 0 | 0 | 0 | 0 |
| 88 | 0 | 0 | 0 | 0 | 0 | 0 |
| 89 | 0 | 0 | 0 | 0 | 0 | 0 |
| 90 | 0 | 0 | 0 | 0 | 0 | 0 |
| 91 | 0 | 0 | 0 | 0 | 0 | 0 |
| 92 | 0 | 0 | 0 | 0 | 0 | 0 |
| 93 | 0 | 0 | 0 | 0 | 0 | 0 |
| 94 | 0 | 0 | 0 | 0 | 0 | 0 |
| 95 | 0 | 0 | 0 | 0 | 0 | 0 |
| 96 | 0 | 0 | 0 | 0 | 0 | 0 |
| 97 | 0 | 0 | 0 | 0 | 0 | 0 |
| 98 | 0 | 0 | 0 | 0 | 0 | 0 |
| 99 | 0 | 0 | 0 | 0 | 0 | 0 |
| 100 | 0 | 0 | 0 | 0 | 0 | 0 |

**Phase II Full-Scale Experiment**
**Test #** 2 **Chamber #** 2
**Pre-ozonated 0.94 μm YG-Microspheres**

| Intensity | Port #1 | Port #2 | Port #3 | Port #4 | Port #5 | Port #b |
|---|---|---|---|---|---|---|
| **1** | 8 | 5 | | 5 | 6 | |
| **2** | 56 | 42 | | 60 | 52 | |
| **3** | 92 | 96 | | 105 | 81 | |
| **4** | 73 | 86 | | 86 | 62 | |
| **5** | 51 | 49 | | 66 | 34 | |
| **6** | 29 | 32 | | 30 | 17 | |
| **7** | 25 | 27 | | 33 | 27 | |
| **8** | 35 | 30 | | 32 | 20 | |
| **9** | 37 | 51 | | 48 | 34 | |
| **10** | 39 | 63 | | 49 | 37 | |
| **11** | 47 | 81 | | 58 | 69 | |
| **12** | 66 | 92 | | 78 | 72 | |
| **13** | 70 | 115 | | 74 | 61 | |
| **14** | 73 | 125 | | 110 | 108 | |
| **15** | 85 | 136 | | 96 | 107 | |
| **16** | 110 | 169 | | 135 | 108 | |
| **17** | 114 | 168 | | 155 | 143 | |
| **18** | 139 | 197 | | 153 | 141 | |
| **19** | 142 | 199 | | 158 | 177 | |
| **20** | 140 | 203 | | 171 | 161 | |
| **21** | 164 | 205 | | 180 | 165 | |
| **22** | 152 | 200 | | 170 | 170 | |
| **23** | 149 | 186 | | 180 | 168 | |
| **24** | 156 | 179 | | 191 | 156 | |
| **25** | 160 | 177 | | 187 | 157 | |
| **26** | 134 | 156 | | 162 | 160 | |
| **27** | 126 | 156 | | 166 | 145 | |
| **28** | 141 | 141 | | 150 | 114 | |
| **29** | 162 | 132 | | 139 | 115 | |
| **30** | 121 | 110 | | 135 | 128 | |
| **31** | 130 | 94 | | 138 | 116 | |
| **32** | 114 | 87 | | 106 | 98 | |
| **33** | 99 | 68 | | 118 | 80 | |
| **34** | 102 | 51 | | 104 | 79 | |
| **35** | 103 | 56 | | 83 | 71 | |
| **36** | 97 | 52 | | 81 | 53 | |
| **37** | 74 | 35 | | 67 | 48 | |
| **38** | 62 | 26 | | 43 | 47 | |
| **39** | 62 | 27 | | 65 | 40 | |
| **40** | 45 | 23 | | 36 | 30 | |
| **41** | 40 | 17 | | 37 | 28 | |
| **42** | 36 | 17 | | 25 | 26 | |
| **43** | 26 | 15 | | 24 | 16 | |
| **44** | 20 | 10 | | 30 | 19 | |
| **45** | 20 | 2 | | 19 | 8 | |
| **46** | 19 | 7 | | 11 | 10 | |
| **47** | 15 | 0 | | 13 | 6 | |
| **48** | 19 | 1 | | 5 | 3 | |
| **49** | 10 | 2 | | 8 | 4 | |
| **50** | 7 | 1 | | 3 | 5 | |

| Intensity | Port #1 | Port #2 | Port #3 | Port #4 | Port #5 | Port #b |
|---|---|---|---|---|---|---|
| **51** | 3 | 0 | | 5 | 3 | |
| **52** | 2 | 1 | | 3 | 1 | |
| **53** | 5 | 0 | | 0 | 1 | |
| **54** | 2 | 0 | | 1 | 1 | |
| **55** | 5 | 0 | | 0 | 0 | |
| **56** | 3 | 0 | | 0 | 0 | |
| **57** | 3 | 0 | | 0 | 1 | |
| **58** | 3 | 0 | | 0 | 0 | |
| **59** | 0 | 0 | | 1 | 0 | |
| **60** | 0 | 0 | | 0 | 0 | |
| **61** | 1 | 0 | | 0 | 0 | |
| **62** | 0 | 0 | | 0 | 0 | |
| **63** | 1 | 0 | | 0 | 0 | |
| **64** | 0 | 0 | | 0 | 0 | |
| **65** | 0 | 0 | | 0 | 0 | |
| **66** | 0 | 0 | | 0 | 0 | |
| **67** | 0 | 0 | | 0 | 0 | |
| **68** | 0 | 0 | | 0 | 0 | |
| **69** | 0 | 0 | | 0 | 0 | |
| **70** | 0 | 0 | | 1 | 0 | |
| **71** | 0 | 0 | | 0 | 0 | |
| **72** | 0 | 0 | | 0 | 0 | |
| **73** | 0 | 0 | | 0 | 0 | |
| **74** | 0 | 0 | | 0 | 0 | |
| **75** | 0 | 0 | | 0 | 0 | |
| **76** | 0 | 0 | | 0 | 0 | |
| **77** | 0 | 0 | | 0 | 0 | |
| **78** | 0 | 0 | | 0 | 0 | |
| **79** | 0 | 0 | | 0 | 0 | |
| **80** | 0 | 0 | | 0 | 0 | |
| **81** | 0 | 0 | | 0 | 0 | |
| **82** | 0 | 0 | | 0 | 0 | |
| **83** | 0 | 0 | | 0 | 0 | |
| **84** | 0 | 0 | | 0 | 0 | |
| **85** | 0 | 0 | | 0 | 0 | |
| **86** | 0 | 0 | | 0 | 0 | |
| **87** | 0 | 0 | | 0 | 0 | |
| **88** | 0 | 0 | | 0 | 0 | |
| **89** | 0 | 0 | | 0 | 0 | |
| **90** | 0 | 0 | | 0 | 0 | |
| **91** | 0 | 0 | | 0 | 0 | |
| **92** | 0 | 0 | | 0 | 0 | |
| **93** | 0 | 0 | | 0 | 0 | |
| **94** | 0 | 0 | | 0 | 0 | |
| **95** | 0 | 0 | | 0 | 0 | |
| **96** | 0 | 0 | | 0 | 0 | |
| **97** | 0 | 0 | | 0 | 0 | |
| **98** | 0 | 0 | | 0 | 0 | |
| **99** | 0 | 0 | | 0 | 0 | |
| **100** | 0 | 0 | | 0 | 0 | |

**Phase II Full-Scale Experiment**
**Test #** 2 **Chamber #** 3
**Pre-ozonated 0.94 $\mu$m YG-Microspheres**

| Intensity | Port #1 | Port #2 | Port #3 | Port #4 | Port #5 | Port #b |
|---|---|---|---|---|---|---|
| **1** | 11 | 10 | 9 | 13 | 8 | 5 |
| **2** | 63 | 75 | 65 | 52 | 69 | 51 |
| **3** | 104 | 131 | 141 | 131 | 124 | 85 |
| **4** | 94 | 142 | 105 | 116 | 102 | 77 |
| **5** | 53 | 74 | 56 | 50 | 53 | 53 |
| **6** | 40 | 50 | 43 | 31 | 53 | 35 |
| **7** | 25 | 44 | 37 | 29 | 28 | 36 |
| **8** | 41 | 33 | 44 | 29 | 27 | 37 |
| **9** | 45 | 55 | 42 | 43 | 43 | 48 |
| **10** | 55 | 57 | 61 | 48 | 53 | 48 |
| **11** | 66 | 84 | 55 | 74 | 60 | 76 |
| **12** | 88 | 88 | 104 | 77 | 83 | 73 |
| **13** | 90 | 105 | 90 | 102 | 80 | 81 |
| **14** | 123 | 119 | 117 | 104 | 88 | 108 |
| **15** | 116 | 147 | 136 | 133 | 113 | 132 |
| **16** | 163 | 161 | 163 | 166 | 120 | 152 |
| **17** | 144 | 175 | 182 | 168 | 130 | 136 |
| **18** | 173 | 194 | 172 | 164 | 168 | 158 |
| **19** | 183 | 226 | 170 | 174 | 134 | 179 |
| **20** | 163 | 234 | 148 | 178 | 154 | 174 |
| **21** | 187 | 205 | 167 | 172 | 165 | 168 |
| **22** | 201 | 224 | 174 | 182 | 170 | 161 |
| **23** | 177 | 222 | 194 | 166 | 150 | 173 |
| **24** | 170 | 203 | 169 | 172 | 137 | 141 |
| **25** | 169 | 183 | 173 | 159 | 120 | 154 |
| **26** | 159 | 197 | 145 | 148 | 142 | 129 |
| **27** | 163 | 156 | 130 | 160 | 122 | 126 |
| **28** | 130 | 165 | 125 | 129 | 107 | 98 |
| **29** | 142 | 144 | 116 | 100 | 90 | 105 |
| **30** | 118 | 125 | 95 | 113 | 78 | 94 |
| **31** | 111 | 132 | 90 | 90 | 76 | 74 |
| **32** | 86 | 94 | 92 | 74 | 65 | 68 |
| **33** | 95 | 69 | 91 | 68 | 51 | 70 |
| **34** | 68 | 82 | 54 | 60 | 50 | 55 |
| **35** | 60 | 56 | 59 | 48 | 24 | 37 |
| **36** | 56 | 50 | 42 | 49 | 32 | 29 |
| **37** | 39 | 43 | 35 | 37 | 23 | 23 |
| **38** | 49 | 30 | 32 | 31 | 19 | 19 |
| **39** | 36 | 24 | 27 | 28 | 15 | 18 |
| **40** | 11 | 22 | 14 | 26 | 13 | 13 |
| **41** | 18 | 21 | 17 | 13 | 12 | 7 |
| **42** | 16 | 14 | 14 | 8 | 7 | 2 |
| **43** | 14 | 13 | 10 | 13 | 5 | 0 |
| **44** | 6 | 5 | 3 | 5 | 6 | 0 |
| **45** | 8 | 7 | 1 | 5 | 0 | 0 |
| **46** | 3 | 1 | 3 | 3 | 3 | 2 |
| **47** | 4 | 0 | 3 | 2 | 0 | 0 |
| **48** | 5 | 3 | 2 | 1 | 0 | 0 |
| **49** | 0 | 2 | 2 | 2 | 0 | 0 |
| **50** | 0 | 0 | 0 | 0 | 0 | 1 |

| Intensity | Port #1 | Port #2 | Port #3 | Port #4 | Port #5 | Port #b |
|---|---|---|---|---|---|---|
| **51** | 1 | 0 | 1 | 0 | 0 | 0 |
| **52** | 2 | 0 | 0 | 0 | 0 | 0 |
| **53** | 0 | 0 | 0 | 0 | 0 | 0 |
| **54** | 0 | 0 | 1 | 0 | 0 | 0 |
| **55** | 0 | 0 | 0 | 0 | 0 | 0 |
| **56** | 0 | 0 | 0 | 0 | 0 | 0 |
| **57** | 0 | 0 | 0 | 1 | 0 | 0 |
| **58** | 0 | 0 | 0 | 0 | 0 | 0 |
| **59** | 0 | 0 | 0 | 0 | 0 | 0 |
| **60** | 1 | 0 | 0 | 0 | 0 | 0 |
| **61** | 0 | 0 | 0 | 0 | 0 | 0 |
| **62** | 1 | 0 | 0 | 0 | 0 | 0 |
| **63** | 0 | 0 | 0 | 0 | 0 | 0 |
| **64** | 0 | 0 | 0 | 0 | 0 | 0 |
| **65** | 0 | 0 | 0 | 0 | 0 | 0 |
| **66** | 0 | 0 | 0 | 0 | 0 | 0 |
| **67** | 0 | 0 | 0 | 0 | 0 | 0 |
| **68** | 0 | 0 | 0 | 0 | 0 | 1 |
| **69** | 0 | 0 | 0 | 0 | 0 | 0 |
| **70** | 0 | 0 | 0 | 0 | 0 | 0 |
| **71** | 0 | 0 | 0 | 0 | 0 | 0 |
| **72** | 0 | 0 | 0 | 0 | 0 | 0 |
| **73** | 0 | 0 | 0 | 0 | 0 | 0 |
| **74** | 0 | 0 | 0 | 0 | 0 | 0 |
| **75** | 0 | 0 | 0 | 0 | 0 | 0 |
| **76** | 0 | 0 | 0 | 0 | 0 | 0 |
| **77** | 0 | 0 | 0 | 0 | 0 | 0 |
| **78** | 0 | 0 | 0 | 0 | 0 | 0 |
| **79** | 0 | 0 | 0 | 0 | 0 | 0 |
| **80** | 0 | 0 | 0 | 0 | 0 | 0 |
| **81** | 0 | 0 | 0 | 0 | 0 | 0 |
| **82** | 0 | 0 | 0 | 0 | 0 | 0 |
| **83** | 0 | 0 | 0 | 0 | 0 | 0 |
| **84** | 0 | 0 | 0 | 0 | 0 | 0 |
| **85** | 0 | 0 | 0 | 0 | 0 | 0 |
| **86** | 0 | 0 | 0 | 0 | 0 | 0 |
| **87** | 0 | 0 | 0 | 0 | 0 | 0 |
| **88** | 0 | 0 | 0 | 0 | 0 | 0 |
| **89** | 0 | 0 | 0 | 0 | 0 | 0 |
| **90** | 0 | 0 | 0 | 0 | 0 | 0 |
| **91** | 0 | 0 | 0 | 0 | 0 | 0 |
| **92** | 0 | 0 | 0 | 0 | 0 | 0 |
| **93** | 0 | 0 | 0 | 0 | 0 | 0 |
| **94** | 0 | 0 | 0 | 0 | 0 | 0 |
| **95** | 0 | 0 | 0 | 0 | 0 | 0 |
| **96** | 0 | 0 | 0 | 0 | 0 | 0 |
| **97** | 0 | 0 | 0 | 0 | 0 | 0 |
| **98** | 0 | 0 | 0 | 0 | 0 | 0 |
| **99** | 0 | 0 | 0 | 0 | 0 | 0 |
| **100** | 0 | 0 | 0 | 0 | 0 | 0 |

**Phase II Full-Scale Experiment**
**Test #** 2 **Chamber #** 4
**Pre-ozonated 0.94 $\mu$m YG-Microspheres**

| Intensity | Port #1 | Port #2 | Port #3 | Port #4 | Port #5 | Port #b |
|---|---|---|---|---|---|---|
| **1** | 11 | | 7 | | 9 | 11 |
| **2** | 53 | | 67 | | 65 | 39 |
| **3** | 96 | | 128 | | 133 | 120 |
| **4** | 91 | | 109 | | 116 | 77 |
| **5** | 54 | | 66 | | 59 | 50 |
| **6** | 38 | | 42 | | 39 | 45 |
| **7** | 40 | | 35 | | 44 | 38 |
| **8** | 46 | | 42 | | 40 | 39 |
| **9** | 46 | | 49 | | 54 | 45 |
| **10** | 62 | | 73 | | 73 | 63 |
| **11** | 68 | | 91 | | 94 | 85 |
| **12** | 78 | | 87 | | 89 | 84 |
| **13** | 108 | | 108 | | 105 | 116 |
| **14** | 107 | | 128 | | 127 | 99 |
| **15** | 139 | | 129 | | 150 | 131 |
| **16** | 144 | | 164 | | 171 | 159 |
| **17** | 165 | | 170 | | 186 | 162 |
| **18** | 178 | | 195 | | 201 | 174 |
| **19** | 204 | | 187 | | 196 | 182 |
| **20** | 175 | | 208 | | 191 | 207 |
| **21** | 198 | | 178 | | 233 | 173 |
| **22** | 166 | | 178 | | 186 | 199 |
| **23** | 186 | | 168 | | 175 | 196 |
| **24** | 178 | | 148 | | 144 | 136 |
| **25** | 144 | | 143 | | 165 | 147 |
| **26** | 135 | | 131 | | 146 | 144 |
| **27** | 122 | | 137 | | 130 | 147 |
| **28** | 111 | | 116 | | 82 | 134 |
| **29** | 114 | | 112 | | 112 | 101 |
| **30** | 108 | | 107 | | 88 | 97 |
| **31** | 72 | | 77 | | 70 | 83 |
| **32** | 82 | | 77 | | 61 | 75 |
| **33** | 70 | | 65 | | 62 | 59 |
| **34** | 77 | | 64 | | 43 | 53 |
| **35** | 39 | | 43 | | 45 | 40 |
| **36** | 36 | | 41 | | 24 | 28 |
| **37** | 32 | | 23 | | 29 | 30 |
| **38** | 13 | | 16 | | 16 | 33 |
| **39** | 16 | | 20 | | 15 | 18 |
| **40** | 12 | | 4 | | 12 | 9 |
| **41** | 5 | | 9 | | 6 | 11 |
| **42** | 3 | | 7 | | 3 | 11 |
| **43** | 8 | | 5 | | 4 | 4 |
| **44** | 2 | | 5 | | 0 | 2 |
| **45** | 5 | | 5 | | 1 | 4 |
| **46** | 3 | | 2 | | 2 | 0 |
| **47** | 1 | | 3 | | 1 | 0 |
| **48** | 2 | | 1 | | 0 | 5 |
| **49** | 2 | | 2 | | 1 | 3 |
| **50** | 0 | | 0 | | 0 | 4 |

| Intensity | Port #1 | Port #2 | Port #3 | Port #4 | Port #5 | Port #b |
|---|---|---|---|---|---|---|
| **51** | 0 | | 0 | | 0 | 1 |
| **52** | 0 | | 0 | | 0 | 2 |
| **53** | 1 | | 0 | | 0 | 0 |
| **54** | 2 | | 0 | | 0 | 0 |
| **55** | 0 | | 0 | | 0 | 3 |
| **56** | 0 | | 0 | | 0 | 1 |
| **57** | 0 | | 0 | | 0 | 1 |
| **58** | 0 | | 0 | | 0 | 1 |
| **59** | 0 | | 0 | | 0 | 1 |
| **60** | 0 | | 0 | | 0 | 0 |
| **61** | 0 | | 0 | | 0 | 0 |
| **62** | 0 | | 0 | | 0 | 1 |
| **63** | 0 | | 1 | | 0 | 0 |
| **64** | 0 | | 0 | | 0 | 1 |
| **65** | 0 | | 1 | | 0 | 0 |
| **66** | 0 | | 0 | | 0 | 1 |
| **67** | 0 | | 0 | | 0 | 0 |
| **68** | 0 | | 0 | | 0 | 0 |
| **69** | 0 | | 0 | | 0 | 3 |
| **70** | 0 | | 0 | | 0 | 0 |
| **71** | 0 | | 0 | | 0 | 0 |
| **72** | 0 | | 0 | | 0 | 2 |
| **73** | 0 | | 0 | | 0 | 1 |
| **74** | 0 | | 0 | | 0 | 0 |
| **75** | 0 | | 0 | | 0 | 0 |
| **76** | 0 | | 0 | | 0 | 0 |
| **77** | 0 | | 0 | | 0 | 0 |
| **78** | 0 | | 0 | | 0 | 0 |
| **79** | 0 | | 0 | | 0 | 0 |
| **80** | 0 | | 0 | | 0 | 1 |
| **81** | 0 | | 0 | | 0 | 0 |
| **82** | 0 | | 0 | | 0 | 0 |
| **83** | 0 | | 0 | | 0 | 0 |
| **84** | 0 | | 0 | | 0 | 0 |
| **85** | 0 | | 0 | | 0 | 0 |
| **86** | 0 | | 0 | | 0 | 0 |
| **87** | 0 | | 0 | | 0 | 0 |
| **88** | 0 | | 0 | | 0 | 0 |
| **89** | 0 | | 0 | | 0 | 0 |
| **90** | 0 | | 0 | | 0 | 0 |
| **91** | 0 | | 0 | | 0 | 0 |
| **92** | 0 | | 0 | | 0 | 0 |
| **93** | 1 | | 0 | | 0 | 0 |
| **94** | 0 | | 0 | | 0 | 0 |
| **95** | 0 | | 0 | | 0 | 0 |
| **96** | 0 | | 1 | | 0 | 0 |
| **97** | 0 | | 0 | | 0 | 0 |
| **98** | 0 | | 0 | | 0 | 0 |
| **99** | 0 | | 0 | | 0 | 0 |
| **100** | 0 | | 0 | | 0 | 0 |

**Phase II Full-Scale Experiment**
**Test #** 2 **Chamber #** 5
**Pre-ozonated 0.94 μm YG-Microspheres**

| Intensity | Port #1 | Port #2 | Port #3 | Port #4 | Port #5 | Port #b |
|---|---|---|---|---|---|---|
| **1** | 5 | | 7 | | 6 | 3 |
| **2** | 34 | | 33 | | 47 | 7 |
| **3** | 79 | | 84 | | 85 | 16 |
| **4** | 65 | | 65 | | 76 | 14 |
| **5** | 33 | | 40 | | 36 | 13 |
| **6** | 23 | | 27 | | 33 | 15 |
| **7** | 21 | | 26 | | 37 | 20 |
| **8** | 31 | | 33 | | 35 | 37 |
| **9** | 38 | | 41 | | 56 | 37 |
| **10** | 69 | | 59 | | 60 | 55 |
| **11** | 61 | | 90 | | 97 | 76 |
| **12** | 77 | | 87 | | 101 | 78 |
| **13** | 101 | | 114 | | 99 | 84 |
| **14** | 102 | | 104 | | 137 | 138 |
| **15** | 126 | | 149 | | 160 | 126 |
| **16** | 153 | | 139 | | 163 | 150 |
| **17** | 183 | | 182 | | 177 | 188 |
| **18** | 169 | | 173 | | 184 | 180 |
| **19** | 212 | | 190 | | 162 | 202 |
| **20** | 179 | | 215 | | 197 | 194 |
| **21** | 173 | | 204 | | 191 | 181 |
| **22** | 161 | | 191 | | 175 | 190 |
| **23** | 217 | | 197 | | 177 | 176 |
| **24** | 154 | | 194 | | 185 | 170 |
| **25** | 169 | | 163 | | 160 | 152 |
| **26** | 142 | | 158 | | 139 | 142 |
| **27** | 143 | | 149 | | 144 | 121 |
| **28** | 120 | | 136 | | 121 | 117 |
| **29** | 101 | | 111 | | 102 | 116 |
| **30** | 91 | | 101 | | 89 | 81 |
| **31** | 87 | | 79 | | 88 | 80 |
| **32** | 78 | | 75 | | 69 | 65 |
| **33** | 57 | | 82 | | 57 | 62 |
| **34** | 37 | | 52 | | 44 | 48 |
| **35** | 60 | | 27 | | 39 | 42 |
| **36** | 30 | | 27 | | 26 | 42 |
| **37** | 31 | | 32 | | 20 | 31 |
| **38** | 15 | | 16 | | 11 | 25 |
| **39** | 11 | | 11 | | 9 | 26 |
| **40** | 15 | | 14 | | 12 | 11 |
| **41** | 12 | | 11 | | 8 | 9 |
| **42** | 7 | | 6 | | 5 | 9 |
| **43** | 5 | | 3 | | 4 | 8 |
| **44** | 6 | | 1 | | 2 | 4 |
| **45** | 3 | | 2 | | 1 | 2 |
| **46** | 2 | | 0 | | 0 | 1 |
| **47** | 1 | | 0 | | 0 | 0 |
| **48** | 3 | | 0 | | 0 | 0 |
| **49** | 0 | | 0 | | 0 | 0 |
| **50** | 1 | | 1 | | 0 | 1 |

| Intensity | Port #1 | Port #2 | Port #3 | Port #4 | Port #5 | Port #b |
|---|---|---|---|---|---|---|
| **51** | 0 | | 0 | | 0 | 0 |
| **52** | 0 | | 0 | | 0 | 0 |
| **53** | 0 | | 0 | | 0 | 0 |
| **54** | 0 | | 0 | | 0 | 1 |
| **55** | 0 | | 0 | | 0 | 0 |
| **56** | 0 | | 0 | | 0 | 0 |
| **57** | 0 | | 0 | | 0 | 0 |
| **58** | 0 | | 0 | | 0 | 0 |
| **59** | 1 | | 0 | | 0 | 0 |
| **60** | 0 | | 0 | | 0 | 0 |
| **61** | 0 | | 0 | | 0 | 0 |
| **62** | 0 | | 0 | | 0 | 0 |
| **63** | 0 | | 0 | | 0 | 0 |
| **64** | 0 | | 0 | | 0 | 0 |
| **65** | 0 | | 0 | | 0 | 0 |
| **66** | 0 | | 0 | | 0 | 0 |
| **67** | 0 | | 0 | | 0 | 0 |
| **68** | 0 | | 0 | | 0 | 0 |
| **69** | 0 | | 0 | | 0 | 0 |
| **70** | 0 | | 0 | | 0 | 0 |
| **71** | 0 | | 0 | | 0 | 0 |
| **72** | 0 | | 0 | | 0 | 0 |
| **73** | 0 | | 0 | | 0 | 0 |
| **74** | 0 | | 0 | | 0 | 0 |
| **75** | 0 | | 0 | | 0 | 0 |
| **76** | 0 | | 0 | | 0 | 0 |
| **77** | 0 | | 0 | | 0 | 0 |
| **78** | 0 | | 0 | | 0 | 0 |
| **79** | 0 | | 0 | | 1 | 0 |
| **80** | 0 | | 0 | | 0 | 0 |
| **81** | 0 | | 0 | | 0 | 0 |
| **82** | 0 | | 0 | | 0 | 0 |
| **83** | 0 | | 0 | | 0 | 0 |
| **84** | 0 | | 0 | | 0 | 0 |
| **85** | 0 | | 0 | | 0 | 0 |
| **86** | 0 | | 0 | | 0 | 0 |
| **87** | 0 | | 0 | | 0 | 0 |
| **88** | 0 | | 0 | | 0 | 0 |
| **89** | 0 | | 0 | | 0 | 0 |
| **90** | 0 | | 0 | | 0 | 0 |
| **91** | 0 | | 0 | | 0 | 0 |
| **92** | 0 | | 0 | | 0 | 0 |
| **93** | 0 | | 0 | | 0 | 0 |
| **94** | 0 | | 0 | | 0 | 0 |
| **95** | 0 | | 0 | | 0 | 0 |
| **96** | 0 | | 0 | | 0 | 0 |
| **97** | 0 | | 0 | | 0 | 0 |
| **98** | 0 | | 0 | | 0 | 0 |
| **99** | 0 | | 0 | | 0 | 0 |
| **100** | 0 | | 0 | | 0 | 0 |

**Phase II Full-Scale Experiment**
**Test #** 3 **Chamber #** 1
**Pre-ozonated 0.94 μm YG-Microspheres**

| Intensity | Port #1 | Port #2 | Port #3 | Port #4 | Port #5 | Port #b |
|---|---|---|---|---|---|---|
| **1** | 7 | 7 | 10 | 9 | 14 | 7 |
| **2** | 50 | 67 | 57 | 56 | 55 | 61 |
| **3** | 122 | 138 | 154 | 108 | 111 | 110 |
| **4** | 83 | 141 | 136 | 114 | 108 | 104 |
| **5** | 64 | 97 | 90 | 72 | 76 | 78 |
| **6** | 42 | 74 | 54 | 48 | 45 | 53 |
| **7** | 20 | 48 | 40 | 26 | 24 | 26 |
| **8** | 8 | 14 | 28 | 22 | 17 | 16 |
| **9** | 7 | 18 | 20 | 18 | 15 | 12 |
| **10** | 8 | 12 | 15 | 17 | 7 | 8 |
| **11** | 5 | 8 | 13 | 11 | 5 | 10 |
| **12** | 11 | 16 | 16 | 11 | 15 | 13 |
| **13** | 7 | 10 | 10 | 19 | 14 | 17 |
| **14** | 12 | 11 | 8 | 19 | 17 | 22 |
| **15** | 9 | 25 | 18 | 18 | 17 | 22 |
| **16** | 15 | 11 | 22 | 27 | 14 | 25 |
| **17** | 24 | 27 | 29 | 18 | 21 | 34 |
| **18** | 20 | 33 | 24 | 47 | 34 | 41 |
| **19** | 26 | 28 | 41 | 42 | 36 | 50 |
| **20** | 27 | 44 | 37 | 51 | 46 | 38 |
| **21** | 27 | 61 | 53 | 61 | 45 | 62 |
| **22** | 54 | 40 | 52 | 59 | 69 | 66 |
| **23** | 40 | 50 | 64 | 73 | 50 | 101 |
| **24** | 53 | 69 | 82 | 82 | 58 | 89 |
| **25** | 51 | 79 | 99 | 87 | 81 | 90 |
| **26** | 64 | 86 | 84 | 94 | 75 | 99 |
| **27** | 67 | 105 | 98 | 108 | 89 | 102 |
| **28** | 70 | 89 | 103 | 107 | 84 | 117 |
| **29** | 70 | 76 | 99 | 101 | 111 | 126 |
| **30** | 75 | 112 | 93 | 108 | 100 | 121 |
| **31** | 85 | 101 | 101 | 114 | 96 | 127 |
| **32** | 77 | 95 | 103 | 110 | 83 | 117 |
| **33** | 89 | 123 | 109 | 134 | 114 | 112 |
| **34** | 67 | 116 | 114 | 120 | 99 | 127 |
| **35** | 92 | 123 | 108 | 136 | 107 | 98 |
| **36** | 98 | 111 | 115 | 132 | 102 | 125 |
| **37** | 90 | 102 | 100 | 93 | 107 | 112 |
| **38** | 88 | 104 | 105 | 121 | 108 | 105 |
| **39** | 94 | 98 | 123 | 104 | 111 | 102 |
| **40** | 95 | 87 | 92 | 84 | 92 | 82 |
| **41** | 81 | 88 | 87 | 101 | 92 | 86 |
| **42** | 81 | 76 | 93 | 89 | 96 | 83 |
| **43** | 80 | 77 | 96 | 85 | 77 | 58 |
| **44** | 74 | 55 | 86 | 79 | 87 | 75 |
| **45** | 71 | 75 | 77 | 93 | 81 | 59 |
| **46** | 71 | 75 | 67 | 70 | 65 | 70 |
| **47** | 64 | 64 | 67 | 70 | 71 | 71 |
| **48** | 83 | 65 | 73 | 56 | 58 | 55 |
| **49** | 53 | 68 | 44 | 55 | 58 | 43 |
| **50** | 77 | 62 | 51 | 55 | 60 | 36 |

| Intensity | Port #1 | Port #2 | Port #3 | Port #4 | Port #5 | Port #b |
|---|---|---|---|---|---|---|
| **51** | 66 | 47 | 47 | 48 | 47 | 31 |
| **52** | 70 | 36 | 37 | 50 | 50 | 28 |
| **53** | 52 | 37 | 34 | 28 | 40 | 24 |
| **54** | 65 | 45 | 30 | 29 | 24 | 16 |
| **55** | 54 | 32 | 25 | 23 | 32 | 19 |
| **56** | 47 | 27 | 16 | 25 | 25 | 16 |
| **57** | 43 | 29 | 16 | 20 | 23 | 6 |
| **58** | 40 | 29 | 13 | 14 | 17 | 9 |
| **59** | 37 | 20 | 14 | 9 | 21 | 6 |
| **60** | 32 | 20 | 9 | 14 | 17 | 10 |
| **61** | 29 | 14 | 11 | 6 | 12 | 3 |
| **62** | 35 | 10 | 7 | 14 | 11 | 3 |
| **63** | 24 | 7 | 8 | 6 | 11 | 1 |
| **64** | 19 | 10 | 6 | 1 | 13 | 2 |
| **65** | 22 | 3 | 3 | 4 | 4 | 1 |
| **66** | 19 | 4 | 2 | 0 | 7 | 1 |
| **67** | 26 | 4 | 3 | 1 | 6 | 0 |
| **68** | 17 | 4 | 1 | 2 | 4 | 1 |
| **69** | 15 | 2 | 1 | 0 | 1 | 2 |
| **70** | 14 | 2 | 3 | 2 | 2 | 0 |
| **71** | 22 | 6 | 0 | 1 | 3 | 0 |
| **72** | 11 | 2 | 0 | 0 | 1 | 0 |
| **73** | 7 | 0 | 0 | 0 | 0 | 0 |
| **74** | 10 | 0 | 1 | 0 | 0 | 0 |
| **75** | 7 | 2 | 1 | 1 | 2 | 0 |
| **76** | 2 | 0 | 1 | 0 | 0 | 0 |
| **77** | 3 | 0 | 0 | 1 | 1 | 0 |
| **78** | 4 | 1 | 0 | 0 | 0 | 0 |
| **79** | 2 | 0 | 0 | 0 | 0 | 0 |
| **80** | 2 | 0 | 0 | 0 | 0 | 0 |
| **81** | 1 | 0 | 1 | 0 | 0 | 0 |
| **82** | 0 | 0 | 0 | 0 | 0 | 0 |
| **83** | 0 | 0 | 0 | 0 | 0 | 0 |
| **84** | 1 | 0 | 0 | 0 | 0 | 0 |
| **85** | 0 | 0 | 0 | 0 | 0 | 0 |
| **86** | 1 | 0 | 0 | 0 | 0 | 0 |
| **87** | 0 | 0 | 0 | 0 | 0 | 0 |
| **88** | 0 | 0 | 0 | 0 | 0 | 0 |
| **89** | 0 | 0 | 0 | 0 | 0 | 0 |
| **90** | 0 | 0 | 0 | 0 | 0 | 0 |
| **91** | 2 | 0 | 0 | 0 | 0 | 0 |
| **92** | 1 | 0 | 0 | 0 | 0 | 0 |
| **93** | 0 | 0 | 0 | 0 | 1 | 0 |
| **94** | 0 | 0 | 0 | 0 | 0 | 0 |
| **95** | 0 | 0 | 0 | 0 | 0 | 0 |
| **96** | 0 | 0 | 0 | 0 | 0 | 0 |
| **97** | 0 | 0 | 0 | 0 | 0 | 0 |
| **98** | 0 | 0 | 0 | 0 | 0 | 0 |
| **99** | 0 | 0 | 0 | 0 | 0 | 0 |
| **100** | 0 | 0 | 0 | 0 | 0 | 0 |

**Phase II Full-Scale Experiment**
**Test #** 3 **Chamber #** 2
**Pre-ozonated 0.94 μm YG-Microspheres**

| Intensity | Port #1 | Port #2 | Port #3 | Port #4 | Port #5 | Port #b |
|---|---|---|---|---|---|---|
| **1** | 10 | 5 | 4 | 10 | 22 | |
| **2** | 66 | 35 | 35 | 63 | 74 | |
| **3** | 109 | 87 | 59 | 149 | 169 | |
| **4** | 95 | 80 | 67 | 125 | 158 | |
| **5** | 59 | 41 | 40 | 66 | 91 | |
| **6** | 44 | 39 | 29 | 37 | 71 | |
| **7** | 23 | 13 | 17 | 30 | 33 | |
| **8** | 15 | 21 | 17 | 24 | 23 | |
| **9** | 12 | 11 | 8 | 13 | 18 | |
| **10** | 7 | 8 | 12 | 14 | 17 | |
| **11** | 5 | 8 | 13 | 17 | 17 | |
| **12** | 19 | 18 | 23 | 18 | 20 | |
| **13** | 14 | 16 | 28 | 15 | 20 | |
| **14** | 23 | 22 | 31 | 17 | 29 | |
| **15** | 16 | 39 | 38 | 19 | 27 | |
| **16** | 35 | 26 | 48 | 42 | 61 | |
| **17** | 39 | 48 | 55 | 40 | 61 | |
| **18** | 33 | 45 | 78 | 52 | 60 | |
| **19** | 46 | 53 | 72 | 63 | 61 | |
| **20** | 60 | 68 | 80 | 61 | 85 | |
| **21** | 83 | 78 | 103 | 84 | 108 | |
| **22** | 84 | 98 | 98 | 88 | 107 | |
| **23** | 100 | 108 | 121 | 113 | 117 | |
| **24** | 109 | 121 | 119 | 91 | 123 | |
| **25** | 103 | 105 | 133 | 113 | 132 | |
| **26** | 130 | 125 | 106 | 150 | 160 | |
| **27** | 121 | 147 | 136 | 128 | 154 | |
| **28** | 97 | 137 | 121 | 143 | 173 | |
| **29** | 136 | 140 | 122 | 142 | 182 | |
| **30** | 131 | 139 | 97 | 138 | 160 | |
| **31** | 116 | 121 | 117 | 117 | 150 | |
| **32** | 106 | 141 | 111 | 140 | 185 | |
| **33** | 136 | 141 | 123 | 115 | 176 | |
| **34** | 97 | 119 | 85 | 108 | 152 | |
| **35** | 126 | 132 | 115 | 138 | 143 | |
| **36** | 114 | 92 | 85 | 106 | 137 | |
| **37** | 110 | 120 | 74 | 110 | 129 | |
| **38** | 100 | 100 | 79 | 90 | 118 | |
| **39** | 91 | 81 | 74 | 90 | 133 | |
| **40** | 96 | 80 | 67 | 88 | 105 | |
| **41** | 82 | 75 | 59 | 72 | 99 | |
| **42** | 78 | 60 | 55 | 72 | 102 | |
| **43** | 70 | 57 | 44 | 74 | 77 | |
| **44** | 58 | 67 | 39 | 49 | 86 | |
| **45** | 69 | 66 | 43 | 61 | 69 | |
| **46** | 64 | 61 | 25 | 55 | 60 | |
| **47** | 45 | 50 | 29 | 46 | 51 | |
| **48** | 37 | 36 | 19 | 30 | 46 | |
| **49** | 41 | 26 | 25 | 31 | 30 | |
| **50** | 33 | 15 | 19 | 23 | 34 | |

| Intensity | Port #1 | Port #2 | Port #3 | Port #4 | Port #5 | Port #b |
|---|---|---|---|---|---|---|
| **51** | 23 | 21 | 15 | 16 | 30 | |
| **52** | 19 | 15 | 10 | 22 | 17 | |
| **53** | 15 | 11 | 15 | 14 | 20 | |
| **54** | 11 | 10 | 9 | 7 | 9 | |
| **55** | 17 | 10 | 4 | 5 | 8 | |
| **56** | 12 | 3 | 4 | 6 | 10 | |
| **57** | 9 | 3 | 5 | 4 | 2 | |
| **58** | 8 | 7 | 2 | 4 | 4 | |
| **59** | 2 | 5 | 3 | 1 | 2 | |
| **60** | 6 | 5 | 3 | 0 | 1 | |
| **61** | 3 | 2 | 1 | 0 | 2 | |
| **62** | 4 | 0 | 0 | 0 | 0 | |
| **63** | 3 | 0 | 2 | 1 | 1 | |
| **64** | 2 | 0 | 0 | 0 | 0 | |
| **65** | 1 | 0 | 0 | 0 | 0 | |
| **66** | 0 | 0 | 1 | 0 | 1 | |
| **67** | 0 | 0 | 0 | 0 | 1 | |
| **68** | 0 | 0 | 0 | 0 | 0 | |
| **69** | 0 | 0 | 0 | 0 | 0 | |
| **70** | 0 | 0 | 1 | 0 | 0 | |
| **71** | 0 | 0 | 0 | 0 | 0 | |
| **72** | 0 | 0 | 1 | 0 | 0 | |
| **73** | 0 | 0 | 0 | 0 | 0 | |
| **74** | 0 | 0 | 0 | 0 | 0 | |
| **75** | 0 | 0 | 0 | 0 | 1 | |
| **76** | 0 | 0 | 0 | 0 | 0 | |
| **77** | 0 | 0 | 0 | 0 | 0 | |
| **78** | 0 | 0 | 0 | 0 | 0 | |
| **79** | 0 | 0 | 0 | 0 | 0 | |
| **80** | 0 | 0 | 0 | 0 | 0 | |
| **81** | 0 | 0 | 0 | 0 | 0 | |
| **82** | 0 | 0 | 0 | 0 | 0 | |
| **83** | 0 | 0 | 0 | 0 | 0 | |
| **84** | 0 | 0 | 0 | 0 | 0 | |
| **85** | 0 | 0 | 0 | 0 | 0 | |
| **86** | 0 | 0 | 0 | 0 | 0 | |
| **87** | 0 | 0 | 0 | 0 | 0 | |
| **88** | 0 | 0 | 0 | 0 | 0 | |
| **89** | 0 | 0 | 0 | 0 | 0 | |
| **90** | 0 | 0 | 0 | 0 | 0 | |
| **91** | 0 | 0 | 0 | 0 | 0 | |
| **92** | 0 | 0 | 0 | 0 | 0 | |
| **93** | 0 | 0 | 0 | 0 | 0 | |
| **94** | 0 | 0 | 0 | 0 | 0 | |
| **95** | 0 | 0 | 0 | 0 | 0 | |
| **96** | 0 | 0 | 0 | 0 | 0 | |
| **97** | 0 | 0 | 0 | 0 | 0 | |
| **98** | 0 | 0 | 0 | 0 | 0 | |
| **99** | 0 | 0 | 0 | 0 | 0 | |
| **100** | 0 | 0 | 0 | 0 | 0 | |

**Phase II Full-Scale Experiment**
**Test #** 3 **Chamber #** 3
**Pre-ozonated 0.94 μm YG-Microspheres**

| Intensity | Port #1 | Port #2 | Port #3 | Port #4 | Port #5 | Port #b |
|---|---|---|---|---|---|---|
| 1 | 15 | | 5 | | 6 | 9 |
| 2 | 70 | | 61 | | 50 | 55 |
| 3 | 109 | | 108 | | 86 | 129 |
| 4 | 101 | | 113 | | 76 | 111 |
| 5 | 64 | | 69 | | 39 | 74 |
| 6 | 38 | | 46 | | 19 | 33 |
| 7 | 18 | | 23 | | 20 | 25 |
| 8 | 18 | | 15 | | 14 | 18 |
| 9 | 11 | | 12 | | 13 | 9 |
| 10 | 14 | | 9 | | 15 | 14 |
| 11 | 21 | | 12 | | 22 | 9 |
| 12 | 20 | | 14 | | 26 | 17 |
| 13 | 32 | | 14 | | 28 | 24 |
| 14 | 44 | | 15 | | 40 | 16 |
| 15 | 34 | | 20 | | 45 | 24 |
| 16 | 49 | | 39 | | 55 | 36 |
| 17 | 52 | | 50 | | 69 | 49 |
| 18 | 56 | | 45 | | 75 | 58 |
| 19 | 72 | | 58 | | 87 | 60 |
| 20 | 73 | | 70 | | 99 | 80 |
| 21 | 100 | | 73 | | 122 | 88 |
| 22 | 105 | | 71 | | 132 | 92 |
| 23 | 126 | | 109 | | 134 | 103 |
| 24 | 123 | | 113 | | 144 | 131 |
| 25 | 138 | | 122 | | 146 | 125 |
| 26 | 147 | | 112 | | 155 | 123 |
| 27 | 157 | | 145 | | 134 | 142 |
| 28 | 144 | | 140 | | 129 | 131 |
| 29 | 148 | | 123 | | 128 | 143 |
| 30 | 139 | | 155 | | 146 | 143 |
| 31 | 139 | | 145 | | 163 | 135 |
| 32 | 106 | | 128 | | 115 | 128 |
| 33 | 129 | | 110 | | 103 | 127 |
| 34 | 119 | | 121 | | 92 | 109 |
| 35 | 96 | | 126 | | 96 | 101 |
| 36 | 105 | | 102 | | 90 | 108 |
| 37 | 84 | | 93 | | 92 | 105 |
| 38 | 66 | | 105 | | 60 | 99 |
| 39 | 84 | | 87 | | 80 | 68 |
| 40 | 73 | | 81 | | 65 | 76 |
| 41 | 58 | | 59 | | 44 | 59 |
| 42 | 50 | | 62 | | 44 | 57 |
| 43 | 41 | | 48 | | 32 | 47 |
| 44 | 33 | | 62 | | 21 | 44 |
| 45 | 32 | | 37 | | 25 | 41 |
| 46 | 27 | | 40 | | 20 | 38 |
| 47 | 22 | | 35 | | 16 | 23 |
| 48 | 19 | | 28 | | 11 | 30 |
| 49 | 22 | | 12 | | 13 | 21 |
| 50 | 13 | | 17 | | 10 | 15 |

| Intensity | Port #1 | Port #2 | Port #3 | Port #4 | Port #5 | Port #b |
|---|---|---|---|---|---|---|
| 51 | 5 | | 15 | | 8 | 9 |
| 52 | 4 | | 10 | | 2 | 11 |
| 53 | 3 | | 18 | | 4 | 4 |
| 54 | 2 | | 5 | | 2 | 5 |
| 55 | 7 | | 2 | | 1 | 3 |
| 56 | 1 | | 1 | | 2 | 9 |
| 57 | 3 | | 3 | | 1 | 0 |
| 58 | 2 | | 1 | | 0 | 1 |
| 59 | 1 | | 3 | | 0 | 0 |
| 60 | 0 | | 1 | | 0 | 0 |
| 61 | 1 | | 0 | | 2 | 0 |
| 62 | 0 | | 0 | | 0 | 0 |
| 63 | 0 | | 0 | | 0 | 1 |
| 64 | 0 | | 0 | | 0 | 0 |
| 65 | 0 | | 0 | | 0 | 0 |
| 66 | 0 | | 0 | | 0 | 0 |
| 67 | 0 | | 0 | | 0 | 0 |
| 68 | 0 | | 0 | | 0 | 0 |
| 69 | 0 | | 0 | | 0 | 0 |
| 70 | 0 | | 0 | | 0 | 0 |
| 71 | 0 | | 0 | | 0 | 0 |
| 72 | 0 | | 0 | | 0 | 0 |
| 73 | 0 | | 0 | | 0 | 0 |
| 74 | 0 | | 0 | | 0 | 0 |
| 75 | 0 | | 0 | | 0 | 0 |
| 76 | 0 | | 0 | | 0 | 0 |
| 77 | 0 | | 0 | | 0 | 0 |
| 78 | 0 | | 0 | | 0 | 0 |
| 79 | 0 | | 0 | | 0 | 0 |
| 80 | 0 | | 0 | | 0 | 0 |
| 81 | 0 | | 0 | | 0 | 0 |
| 82 | 0 | | 0 | | 0 | 0 |
| 83 | 0 | | 0 | | 0 | 0 |
| 84 | 0 | | 0 | | 0 | 0 |
| 85 | 0 | | 0 | | 0 | 0 |
| 86 | 0 | | 0 | | 0 | 0 |
| 87 | 0 | | 0 | | 0 | 0 |
| 88 | 0 | | 0 | | 0 | 0 |
| 89 | 0 | | 0 | | 0 | 0 |
| 90 | 0 | | 0 | | 0 | 0 |
| 91 | 0 | | 0 | | 0 | 0 |
| 92 | 0 | | 0 | | 0 | 0 |
| 93 | 0 | | 0 | | 0 | 0 |
| 94 | 0 | | 0 | | 0 | 0 |
| 95 | 0 | | 0 | | 0 | 0 |
| 96 | 0 | | 0 | | 0 | 0 |
| 97 | 0 | | 0 | | 0 | 0 |
| 98 | 0 | | 0 | | 0 | 0 |
| 99 | 0 | | 0 | | 0 | 0 |
| 100 | 0 | | 0 | | 0 | 0 |

**Phase II Full-Scale Experiment**
**Test #** 3 **Chamber #** 4
**Pre-ozonated 0.94 μm YG-Microspheres**

| Intensity | Port #1 | Port #2 | Port #3 | Port #4 | Port #5 | Port #b |
|---|---|---|---|---|---|---|
| **1** | 11 | | 9 | | 9 | 10 |
| **2** | 61 | | 57 | | 54 | 60 |
| **3** | 141 | | 100 | | 122 | 116 |
| **4** | 134 | | 99 | | 118 | 119 |
| **5** | 94 | | 74 | | 68 | 80 |
| **6** | 55 | | 53 | | 46 | 37 |
| **7** | 43 | | 26 | | 18 | 17 |
| **8** | 25 | | 17 | | 15 | 17 |
| **9** | 13 | | 14 | | 14 | 20 |
| **10** | 20 | | 17 | | 18 | 16 |
| **11** | 9 | | 12 | | 16 | 16 |
| **12** | 18 | | 15 | | 17 | 27 |
| **13** | 16 | | 19 | | 21 | 26 |
| **14** | 25 | | 23 | | 14 | 25 |
| **15** | 32 | | 27 | | 18 | 41 |
| **16** | 38 | | 39 | | 35 | 46 |
| **17** | 52 | | 51 | | 47 | 48 |
| **18** | 57 | | 66 | | 59 | 76 |
| **19** | 81 | | 78 | | 66 | 79 |
| **20** | 78 | | 89 | | 75 | 107 |
| **21** | 73 | | 112 | | 63 | 113 |
| **22** | 105 | | 100 | | 93 | 128 |
| **23** | 116 | | 131 | | 109 | 119 |
| **24** | 123 | | 114 | | 109 | 151 |
| **25** | 123 | | 131 | | 116 | 142 |
| **26** | 142 | | 137 | | 120 | 161 |
| **27** | 158 | | 166 | | 135 | 143 |
| **28** | 149 | | 128 | | 120 | 143 |
| **29** | 144 | | 117 | | 137 | 143 |
| **30** | 134 | | 140 | | 137 | 151 |
| **31** | 145 | | 135 | | 148 | 146 |
| **32** | 131 | | 117 | | 123 | 151 |
| **33** | 116 | | 140 | | 121 | 136 |
| **34** | 131 | | 116 | | 115 | 108 |
| **35** | 114 | | 123 | | 110 | 98 |
| **36** | 101 | | 83 | | 114 | 86 |
| **37** | 91 | | 89 | | 101 | 92 |
| **38** | 81 | | 84 | | 94 | 79 |
| **39** | 96 | | 84 | | 87 | 63 |
| **40** | 79 | | 64 | | 70 | 81 |
| **41** | 73 | | 69 | | 75 | 47 |
| **42** | 68 | | 54 | | 61 | 55 |
| **43** | 60 | | 52 | | 54 | 33 |
| **44** | 45 | | 36 | | 53 | 32 |
| **45** | 41 | | 31 | | 39 | 31 |
| **46** | 32 | | 24 | | 47 | 17 |
| **47** | 32 | | 27 | | 40 | 29 |
| **48** | 28 | | 25 | | 32 | 21 |
| **49** | 26 | | 24 | | 23 | 13 |
| **50** | 27 | | 10 | | 24 | 14 |

| Intensity | Port #1 | Port #2 | Port #3 | Port #4 | Port #5 | Port #b |
|---|---|---|---|---|---|---|
| **51** | 15 | | 15 | | 11 | 10 |
| **52** | 7 | | 9 | | 13 | 5 |
| **53** | 9 | | 6 | | 12 | 5 |
| **54** | 5 | | 0 | | 11 | 6 |
| **55** | 6 | | 2 | | 9 | 4 |
| **56** | 6 | | 3 | | 3 | 0 |
| **57** | 4 | | 4 | | 5 | 1 |
| **58** | 2 | | 1 | | 5 | 1 |
| **59** | 0 | | 1 | | 2 | 1 |
| **60** | 0 | | 1 | | 3 | 2 |
| **61** | 2 | | 0 | | 0 | 0 |
| **62** | 0 | | 0 | | 0 | 0 |
| **63** | 0 | | 0 | | 0 | 0 |
| **64** | 0 | | 0 | | 0 | 0 |
| **65** | 0 | | 0 | | 0 | 0 |
| **66** | 0 | | 0 | | 0 | 0 |
| **67** | 0 | | 0 | | 0 | 0 |
| **68** | 0 | | 0 | | 1 | 0 |
| **69** | 0 | | 0 | | 0 | 0 |
| **70** | 0 | | 0 | | 0 | 0 |
| **71** | 0 | | 0 | | 0 | 0 |
| **72** | 0 | | 0 | | 0 | 0 |
| **73** | 0 | | 0 | | 0 | 0 |
| **74** | 0 | | 0 | | 0 | 0 |
| **75** | 0 | | 0 | | 0 | 0 |
| **76** | 0 | | 0 | | 0 | 0 |
| **77** | 0 | | 0 | | 0 | 0 |
| **78** | 0 | | 0 | | 0 | 0 |
| **79** | 0 | | 0 | | 0 | 0 |
| **80** | 0 | | 0 | | 0 | 0 |
| **81** | 0 | | 0 | | 0 | 0 |
| **82** | 0 | | 0 | | 0 | 0 |
| **83** | 0 | | 0 | | 0 | 0 |
| **84** | 0 | | 0 | | 0 | 0 |
| **85** | 0 | | 0 | | 0 | 0 |
| **86** | 0 | | 0 | | 0 | 0 |
| **87** | 0 | | 0 | | 0 | 0 |
| **88** | 0 | | 0 | | 0 | 0 |
| **89** | 0 | | 0 | | 0 | 0 |
| **90** | 0 | | 0 | | 0 | 0 |
| **91** | 0 | | 0 | | 0 | 0 |
| **92** | 0 | | 0 | | 0 | 0 |
| **93** | 0 | | 0 | | 0 | 0 |
| **94** | 0 | | 0 | | 0 | 0 |
| **95** | 0 | | 0 | | 0 | 0 |
| **96** | 0 | | 0 | | 0 | 0 |
| **97** | 0 | | 0 | | 0 | 0 |
| **98** | 0 | | 0 | | 0 | 0 |
| **99** | 0 | | 0 | | 0 | 0 |
| **100** | 0 | | 0 | | 0 | 0 |

**Phase II Full-Scale Experiment**
**Test #** 3 **Chamber #** 5
**Pre-ozonated 0.94 μm YG-Microspheres**

| Intensity | Port #1 | Port #2 | Port #3 | Port #4 | Port #5 | Port #b |
|---|---|---|---|---|---|---|
| 1 | 6 | | 1 | | 5 | 9 |
| 2 | 15 | | 16 | | 63 | 61 |
| 3 | 65 | | 34 | | 99 | 125 |
| 4 | 48 | | 24 | | 102 | 91 |
| 5 | 32 | | 11 | | 67 | 54 |
| 6 | 23 | | 11 | | 30 | 31 |
| 7 | 15 | | 7 | | 20 | 24 |
| 8 | 10 | | 8 | | 15 | 8 |
| 9 | 10 | | 5 | | 10 | 13 |
| 10 | 10 | | 7 | | 10 | 8 |
| 11 | 19 | | 11 | | 10 | 20 |
| 12 | 17 | | 14 | | 22 | 17 |
| 13 | 22 | | 15 | | 13 | 29 |
| 14 | 32 | | 25 | | 20 | 34 |
| 15 | 50 | | 45 | | 29 | 33 |
| 16 | 45 | | 37 | | 27 | 57 |
| 17 | 56 | | 59 | | 34 | 68 |
| 18 | 60 | | 80 | | 55 | 64 |
| 19 | 79 | | 99 | | 51 | 99 |
| 20 | 108 | | 106 | | 70 | 104 |
| 21 | 118 | | 118 | | 90 | 111 |
| 22 | 108 | | 136 | | 93 | 116 |
| 23 | 100 | | 127 | | 91 | 144 |
| 24 | 136 | | 149 | | 96 | 168 |
| 25 | 137 | | 136 | | 124 | 153 |
| 26 | 142 | | 138 | | 115 | 178 |
| 27 | 149 | | 140 | | 122 | 155 |
| 28 | 154 | | 147 | | 140 | 162 |
| 29 | 121 | | 150 | | 157 | 167 |
| 30 | 146 | | 136 | | 140 | 157 |
| 31 | 140 | | 121 | | 138 | 137 |
| 32 | 112 | | 112 | | 120 | 130 |
| 33 | 139 | | 142 | | 129 | 123 |
| 34 | 121 | | 104 | | 111 | 131 |
| 35 | 106 | | 123 | | 105 | 102 |
| 36 | 116 | | 94 | | 106 | 89 |
| 37 | 86 | | 79 | | 86 | 83 |
| 38 | 90 | | 77 | | 114 | 97 |
| 39 | 84 | | 81 | | 82 | 70 |
| 40 | 62 | | 57 | | 77 | 71 |
| 41 | 77 | | 65 | | 83 | 56 |
| 42 | 59 | | 59 | | 72 | 46 |
| 43 | 46 | | 48 | | 67 | 33 |
| 44 | 47 | | 41 | | 49 | 33 |
| 45 | 30 | | 37 | | 50 | 22 |
| 46 | 27 | | 21 | | 41 | 24 |
| 47 | 15 | | 18 | | 37 | 23 |
| 48 | 15 | | 15 | | 37 | 10 |
| 49 | 16 | | 10 | | 21 | 16 |
| 50 | 11 | | 8 | | 26 | 9 |

| Intensity | Port #1 | Port #2 | Port #3 | Port #4 | Port #5 | Port #b |
|---|---|---|---|---|---|---|
| 51 | 13 | | 16 | | 20 | 11 |
| 52 | 11 | | 3 | | 12 | 4 |
| 53 | 4 | | 4 | | 13 | 3 |
| 54 | 2 | | 1 | | 14 | 2 |
| 55 | 5 | | 4 | | 3 | 1 |
| 56 | 1 | | 2 | | 7 | 0 |
| 57 | 1 | | 1 | | 2 | 0 |
| 58 | 1 | | 0 | | 2 | 0 |
| 59 | 0 | | 1 | | 3 | 1 |
| 60 | 1 | | 0 | | 0 | 1 |
| 61 | 0 | | 0 | | 3 | 0 |
| 62 | 0 | | 1 | | 1 | 0 |
| 63 | 0 | | 0 | | 0 | 0 |
| 64 | 0 | | 0 | | 1 | 0 |
| 65 | 0 | | 0 | | 0 | 0 |
| 66 | 0 | | 0 | | 0 | 1 |
| 67 | 0 | | 0 | | 0 | 0 |
| 68 | 0 | | 0 | | 0 | 0 |
| 69 | 0 | | 0 | | 1 | 0 |
| 70 | 0 | | 0 | | 0 | 0 |
| 71 | 1 | | 0 | | 0 | 0 |
| 72 | 0 | | 0 | | 0 | 0 |
| 73 | 0 | | 0 | | 0 | 0 |
| 74 | 0 | | 0 | | 0 | 0 |
| 75 | 0 | | 0 | | 0 | 0 |
| 76 | 0 | | 0 | | 0 | 0 |
| 77 | 0 | | 0 | | 0 | 0 |
| 78 | 0 | | 0 | | 0 | 0 |
| 79 | 0 | | 0 | | 0 | 0 |
| 80 | 0 | | 0 | | 0 | 0 |
| 81 | 0 | | 0 | | 0 | 0 |
| 82 | 0 | | 0 | | 0 | 0 |
| 83 | 0 | | 0 | | 0 | 0 |
| 84 | 0 | | 0 | | 0 | 0 |
| 85 | 0 | | 0 | | 0 | 0 |
| 86 | 0 | | 0 | | 0 | 0 |
| 87 | 0 | | 0 | | 0 | 0 |
| 88 | 0 | | 0 | | 0 | 0 |
| 89 | 0 | | 0 | | 0 | 0 |
| 90 | 0 | | 0 | | 0 | 0 |
| 91 | 0 | | 0 | | 0 | 0 |
| 92 | 0 | | 0 | | 0 | 0 |
| 93 | 0 | | 0 | | 0 | 0 |
| 94 | 0 | | 0 | | 0 | 0 |
| 95 | 0 | | 0 | | 0 | 0 |
| 96 | 0 | | 0 | | 0 | 0 |
| 97 | 0 | | 0 | | 0 | 0 |
| 98 | 0 | | 0 | | 0 | 0 |
| 99 | 0 | | 0 | | 0 | 0 |
| 100 | 0 | | 0 | | 0 | 0 |

**Phase II Full-Scale Experiment**
**Test #** 4 **Chamber #** 1
**Pre-ozonated 0.94 μm YG-Microspheres**

| Intensity | Port #1 | Port #2 | Port #3 | Port #4 | Port #5 | Port #b |
|---|---|---|---|---|---|---|
| **1** | 10 | 13 | 14 | 12 | 14 | 12 |
| **2** | 77 | 85 | 87 | 82 | 76 | 112 |
| **3** | 154 | 187 | 200 | 183 | 142 | 196 |
| **4** | 133 | 220 | 150 | 151 | 149 | 204 |
| **5** | 93 | 138 | 95 | 87 | 91 | 137 |
| **6** | 51 | 99 | 48 | 66 | 49 | 97 |
| **7** | 26 | 51 | 34 | 44 | 25 | 39 |
| **8** | 23 | 35 | 26 | 24 | 21 | 31 |
| **9** | 9 | 19 | 16 | 21 | 12 | 20 |
| **10** | 7 | 21 | 8 | 7 | 8 | 10 |
| **11** | 1 | 8 | 14 | 11 | 7 | 17 |
| **12** | 0 | 12 | 7 | 8 | 8 | 12 |
| **13** | 6 | 16 | 4 | 13 | 8 | 13 |
| **14** | 7 | 10 | 11 | 15 | 9 | 12 |
| **15** | 6 | 17 | 3 | 16 | 7 | 20 |
| **16** | 4 | 21 | 17 | 13 | 10 | 21 |
| **17** | 12 | 15 | 27 | 27 | 8 | 24 |
| **18** | 12 | 20 | 20 | 22 | 12 | 23 |
| **19** | 16 | 32 | 26 | 32 | 19 | 51 |
| **20** | 16 | 42 | 24 | 42 | 24 | 39 |
| **21** | 14 | 35 | 34 | 59 | 28 | 54 |
| **22** | 31 | 38 | 33 | 49 | 24 | 56 |
| **23** | 40 | 43 | 45 | 60 | 46 | 64 |
| **24** | 29 | 71 | 55 | 72 | 41 | 75 |
| **25** | 26 | 72 | 55 | 74 | 41 | 91 |
| **26** | 45 | 69 | 78 | 86 | 49 | 95 |
| **27** | 54 | 75 | 73 | 82 | 54 | 113 |
| **28** | 41 | 73 | 91 | 109 | 67 | 118 |
| **29** | 52 | 89 | 86 | 80 | 70 | 119 |
| **30** | 62 | 110 | 90 | 92 | 76 | 100 |
| **31** | 56 | 108 | 77 | 115 | 78 | 138 |
| **32** | 54 | 120 | 88 | 123 | 78 | 123 |
| **33** | 56 | 105 | 91 | 135 | 76 | 148 |
| **34** | 68 | 101 | 100 | 118 | 76 | 156 |
| **35** | 90 | 96 | 128 | 118 | 88 | 136 |
| **36** | 74 | 95 | 96 | 127 | 94 | 128 |
| **37** | 79 | 133 | 106 | 114 | 93 | 139 |
| **38** | 69 | 100 | 105 | 122 | 95 | 125 |
| **39** | 76 | 116 | 90 | 110 | 88 | 139 |
| **40** | 82 | 93 | 99 | 104 | 93 | 145 |
| **41** | 79 | 95 | 107 | 101 | 90 | 149 |
| **42** | 94 | 88 | 89 | 94 | 97 | 121 |
| **43** | 79 | 98 | 104 | 106 | 85 | 120 |
| **44** | 82 | 87 | 92 | 99 | 111 | 148 |
| **45** | 76 | 78 | 92 | 81 | 94 | 104 |
| **46** | 73 | 77 | 78 | 66 | 99 | 101 |
| **47** | 70 | 90 | 79 | 68 | 78 | 99 |
| **48** | 67 | 66 | 74 | 64 | 90 | 82 |
| **49** | 75 | 72 | 64 | 62 | 89 | 72 |
| **50** | 90 | 67 | 65 | 54 | 79 | 83 |

| Intensity | Port #1 | Port #2 | Port #3 | Port #4 | Port #5 | Port #b |
|---|---|---|---|---|---|---|
| **51** | 93 | 68 | 60 | 58 | 68 | 93 |
| **52** | 64 | 54 | 49 | 51 | 69 | 75 |
| **53** | 75 | 47 | 50 | 48 | 83 | 42 |
| **54** | 75 | 28 | 49 | 34 | 63 | 40 |
| **55** | 65 | 46 | 57 | 34 | 59 | 57 |
| **56** | 71 | 43 | 41 | 23 | 53 | 38 |
| **57** | 57 | 29 | 40 | 27 | 58 | 40 |
| **58** | 62 | 22 | 29 | 21 | 52 | 35 |
| **59** | 59 | 20 | 26 | 21 | 52 | 28 |
| **60** | 60 | 22 | 20 | 16 | 45 | 28 |
| **61** | 57 | 19 | 25 | 14 | 28 | 17 |
| **62** | 54 | 12 | 19 | 5 | 31 | 18 |
| **63** | 45 | 10 | 20 | 10 | 21 | 8 |
| **64** | 40 | 11 | 11 | 8 | 26 | 12 |
| **65** | 56 | 5 | 18 | 5 | 15 | 8 |
| **66** | 33 | 8 | 12 | 2 | 16 | 4 |
| **67** | 37 | 4 | 10 | 11 | 13 | 4 |
| **68** | 45 | 4 | 5 | 2 | 13 | 3 |
| **69** | 28 | 7 | 7 | 3 | 10 | 4 |
| **70** | 21 | 2 | 3 | 2 | 6 | 3 |
| **71** | 25 | 0 | 4 | 0 | 15 | 2 |
| **72** | 26 | 1 | 2 | 0 | 8 | 2 |
| **73** | 16 | 2 | 0 | 0 | 4 | 0 |
| **74** | 12 | 0 | 0 | 1 | 2 | 0 |
| **75** | 15 | 1 | 1 | 1 | 4 | 0 |
| **76** | 9 | 0 | 0 | 0 | 1 | 0 |
| **77** | 11 | 0 | 1 | 0 | 2 | 0 |
| **78** | 11 | 0 | 2 | 0 | 2 | 0 |
| **79** | 10 | 0 | 1 | 0 | 1 | 0 |
| **80** | 5 | 0 | 0 | 0 | 3 | 0 |
| **81** | 8 | 0 | 0 | 0 | 0 | 0 |
| **82** | 2 | 0 | 0 | 0 | 0 | 0 |
| **83** | 4 | 0 | 0 | 0 | 0 | 0 |
| **84** | 4 | 0 | 0 | 0 | 0 | 0 |
| **85** | 4 | 0 | 0 | 0 | 0 | 0 |
| **86** | 1 | 0 | 0 | 0 | 0 | 0 |
| **87** | 0 | 0 | 0 | 0 | 0 | 0 |
| **88** | 1 | 0 | 0 | 0 | 0 | 0 |
| **89** | 0 | 0 | 0 | 0 | 0 | 0 |
| **90** | 0 | 0 | 0 | 0 | 0 | 0 |
| **91** | 0 | 0 | 0 | 0 | 1 | 0 |
| **92** | 0 | 0 | 0 | 0 | 0 | 0 |
| **93** | 1 | 0 | 0 | 0 | 0 | 0 |
| **94** | 0 | 0 | 0 | 0 | 0 | 0 |
| **95** | 0 | 0 | 0 | 0 | 0 | 0 |
| **96** | 0 | 0 | 0 | 0 | 0 | 0 |
| **97** | 0 | 0 | 0 | 0 | 0 | 0 |
| **98** | 0 | 0 | 0 | 0 | 0 | 0 |
| **99** | 0 | 0 | 0 | 0 | 0 | 0 |
| **100** | 0 | 1 | 0 | 0 | 0 | 0 |

**Phase II Full-Scale Experiment**
**Test #** 4 **Chamber #** 2
**Pre-ozonated 0.94 μm YG-Microspheres**

| Intensity | Port #1 | Port #2 | Port #3 | Port #4 | Port #5 | Port #b |
|---|---|---|---|---|---|---|
| 1 | 20 | 10 | 17 | 18 | 19 | |
| 2 | 85 | 96 | 76 | 70 | 68 | |
| 3 | 174 | 164 | 147 | 125 | 138 | |
| 4 | 169 | 175 | 131 | 125 | 116 | |
| 5 | 131 | 106 | 73 | 62 | 73 | |
| 6 | 74 | 77 | 50 | 37 | 48 | |
| 7 | 40 | 27 | 38 | 20 | 24 | |
| 8 | 28 | 25 | 16 | 14 | 7 | |
| 9 | 25 | 17 | 13 | 15 | 7 | |
| 10 | 10 | 10 | 11 | 13 | 9 | |
| 11 | 15 | 12 | 10 | 4 | 6 | |
| 12 | 10 | 7 | 15 | 10 | 15 | |
| 13 | 14 | 16 | 13 | 11 | 11 | |
| 14 | 14 | 14 | 14 | 15 | 16 | |
| 15 | 18 | 18 | 18 | 23 | 22 | |
| 16 | 16 | 14 | 24 | 15 | 26 | |
| 17 | 33 | 33 | 32 | 34 | 31 | |
| 18 | 30 | 26 | 36 | 35 | 34 | |
| 19 | 37 | 41 | 47 | 36 | 33 | |
| 20 | 41 | 42 | 44 | 40 | 47 | |
| 21 | 45 | 40 | 62 | 62 | 56 | |
| 22 | 61 | 74 | 77 | 64 | 61 | |
| 23 | 81 | 63 | 80 | 76 | 68 | |
| 24 | 72 | 70 | 89 | 86 | 83 | |
| 25 | 69 | 99 | 93 | 85 | 80 | |
| 26 | 90 | 108 | 109 | 102 | 115 | |
| 27 | 96 | 102 | 103 | 106 | 95 | |
| 28 | 108 | 113 | 122 | 116 | 95 | |
| 29 | 120 | 116 | 123 | 120 | 104 | |
| 30 | 133 | 127 | 118 | 148 | 109 | |
| 31 | 126 | 119 | 136 | 112 | 121 | |
| 32 | 136 | 138 | 123 | 134 | 124 | |
| 33 | 105 | 117 | 126 | 116 | 138 | |
| 34 | 134 | 110 | 119 | 110 | 125 | |
| 35 | 107 | 133 | 130 | 107 | 126 | |
| 36 | 114 | 85 | 112 | 124 | 116 | |
| 37 | 123 | 135 | 109 | 113 | 118 | |
| 38 | 119 | 109 | 119 | 109 | 116 | |
| 39 | 95 | 104 | 100 | 114 | 98 | |
| 40 | 95 | 99 | 110 | 105 | 101 | |
| 41 | 96 | 101 | 102 | 94 | 105 | |
| 42 | 83 | 97 | 104 | 88 | 93 | |
| 43 | 68 | 84 | 90 | 76 | 79 | |
| 44 | 62 | 65 | 77 | 59 | 60 | |
| 45 | 73 | 62 | 73 | 59 | 86 | |
| 46 | 64 | 57 | 62 | 71 | 58 | |
| 47 | 56 | 69 | 62 | 52 | 44 | |
| 48 | 67 | 42 | 60 | 47 | 56 | |
| 49 | 51 | 47 | 50 | 42 | 44 | |
| 50 | 41 | 46 | 37 | 38 | 33 | |

| Intensity | Port #1 | Port #2 | Port #3 | Port #4 | Port #5 | Port #b |
|---|---|---|---|---|---|---|
| 51 | 39 | 29 | 45 | 33 | 42 | |
| 52 | 20 | 30 | 23 | 36 | 32 | |
| 53 | 33 | 32 | 27 | 20 | 34 | |
| 54 | 18 | 18 | 28 | 24 | 14 | |
| 55 | 18 | 19 | 21 | 17 | 22 | |
| 56 | 13 | 19 | 17 | 13 | 13 | |
| 57 | 22 | 16 | 17 | 11 | 16 | |
| 58 | 18 | 11 | 16 | 7 | 17 | |
| 59 | 8 | 9 | 10 | 7 | 12 | |
| 60 | 6 | 4 | 5 | 5 | 5 | |
| 61 | 4 | 8 | 4 | 1 | 4 | |
| 62 | 10 | 3 | 5 | 1 | 3 | |
| 63 | 2 | 2 | 4 | 0 | 7 | |
| 64 | 3 | 0 | 5 | 2 | 1 | |
| 65 | 4 | 2 | 2 | 1 | 4 | |
| 66 | 4 | 0 | 0 | 2 | 1 | |
| 67 | 0 | 1 | 0 | 0 | 2 | |
| 68 | 2 | 1 | 2 | 0 | 0 | |
| 69 | 0 | 0 | 0 | 0 | 0 | |
| 70 | 0 | 1 | 0 | 0 | 0 | |
| 71 | 0 | 0 | 0 | 0 | 0 | |
| 72 | 0 | 1 | 0 | 0 | 0 | |
| 73 | 0 | 0 | 0 | 0 | 0 | |
| 74 | 0 | 0 | 0 | 0 | 0 | |
| 75 | 0 | 0 | 0 | 0 | 0 | |
| 76 | 0 | 0 | 0 | 0 | 0 | |
| 77 | 1 | 0 | 0 | 0 | 0 | |
| 78 | 0 | 1 | 0 | 0 | 0 | |
| 79 | 0 | 0 | 0 | 0 | 0 | |
| 80 | 0 | 0 | 0 | 0 | 0 | |
| 81 | 0 | 0 | 0 | 0 | 0 | |
| 82 | 0 | 0 | 0 | 0 | 0 | |
| 83 | 0 | 0 | 0 | 0 | 0 | |
| 84 | 0 | 0 | 0 | 0 | 0 | |
| 85 | 0 | 0 | 0 | 0 | 0 | |
| 86 | 0 | 0 | 0 | 0 | 0 | |
| 87 | 0 | 0 | 0 | 0 | 0 | |
| 88 | 0 | 0 | 0 | 0 | 0 | |
| 89 | 0 | 0 | 0 | 0 | 0 | |
| 90 | 0 | 0 | 0 | 0 | 0 | |
| 91 | 0 | 0 | 0 | 0 | 0 | |
| 92 | 0 | 0 | 0 | 0 | 0 | |
| 93 | 0 | 0 | 0 | 0 | 0 | |
| 94 | 0 | 0 | 0 | 0 | 0 | |
| 95 | 0 | 0 | 0 | 0 | 0 | |
| 96 | 0 | 0 | 0 | 0 | 0 | |
| 97 | 0 | 0 | 0 | 0 | 0 | |
| 98 | 0 | 0 | 0 | 0 | 0 | |
| 99 | 0 | 0 | 0 | 0 | 0 | |
| 100 | 0 | 0 | 0 | 0 | 0 | |

**Phase II Full-Scale Experiment**
**Test #** 4 **Chamber #** 3
**Pre-ozonated 0.94 $\mu$m YG-Microspheres**

| Intensity | Port #1 | Port #2 | Port #3 | Port #4 | Port #5 | Port #b |
|---|---|---|---|---|---|---|
| **1** | 6 | 7 | 6 | 15 | 12 | 19 |
| **2** | 86 | 68 | 63 | 79 | 50 | 93 |
| **3** | 162 | 133 | 116 | 150 | 166 | 228 |
| **4** | 144 | 121 | 92 | 140 | 127 | 153 |
| **5** | 79 | 71 | 57 | 63 | 90 | 119 |
| **6** | 51 | 48 | 28 | 27 | 52 | 67 |
| **7** | 32 | 18 | 21 | 33 | 25 | 40 |
| **8** | 18 | 14 | 15 | 16 | 10 | 18 |
| **9** | 15 | 11 | 11 | 12 | 18 | 14 |
| **10** | 11 | 8 | 10 | 14 | 10 | 8 |
| **11** | 7 | 10 | 3 | 12 | 8 | 10 |
| **12** | 7 | 6 | 4 | 10 | 13 | 14 |
| **13** | 10 | 16 | 15 | 12 | 11 | 7 |
| **14** | 11 | 14 | 15 | 19 | 9 | 19 |
| **15** | 18 | 19 | 15 | 21 | 23 | 18 |
| **16** | 25 | 18 | 25 | 22 | 21 | 33 |
| **17** | 24 | 30 | 19 | 32 | 30 | 46 |
| **18** | 34 | 31 | 24 | 34 | 29 | 34 |
| **19** | 41 | 42 | 43 | 56 | 48 | 45 |
| **20** | 54 | 42 | 48 | 52 | 50 | 63 |
| **21** | 49 | 46 | 64 | 57 | 64 | 60 |
| **22** | 59 | 75 | 61 | 66 | 83 | 79 |
| **23** | 74 | 60 | 79 | 84 | 95 | 74 |
| **24** | 90 | 84 | 92 | 69 | 89 | 86 |
| **25** | 97 | 89 | 100 | 95 | 96 | 130 |
| **26** | 99 | 104 | 100 | 102 | 113 | 115 |
| **27** | 116 | 101 | 103 | 125 | 109 | 124 |
| **28** | 112 | 109 | 94 | 132 | 141 | 126 |
| **29** | 125 | 115 | 110 | 125 | 138 | 137 |
| **30** | 118 | 107 | 126 | 127 | 127 | 134 |
| **31** | 122 | 140 | 128 | 139 | 143 | 120 |
| **32** | 111 | 112 | 128 | 116 | 135 | 110 |
| **33** | 147 | 130 | 137 | 134 | 131 | 119 |
| **34** | 134 | 141 | 145 | 129 | 144 | 139 |
| **35** | 134 | 118 | 132 | 117 | 118 | 138 |
| **36** | 133 | 110 | 109 | 99 | 146 | 111 |
| **37** | 92 | 123 | 100 | 121 | 122 | 112 |
| **38** | 115 | 106 | 105 | 118 | 105 | 121 |
| **39** | 90 | 110 | 104 | 107 | 119 | 120 |
| **40** | 117 | 89 | 107 | 99 | 85 | 75 |
| **41** | 59 | 96 | 90 | 91 | 86 | 75 |
| **42** | 74 | 87 | 87 | 92 | 77 | 90 |
| **43** | 86 | 75 | 77 | 74 | 74 | 64 |
| **44** | 61 | 87 | 67 | 67 | 70 | 78 |
| **45** | 73 | 61 | 64 | 58 | 65 | 67 |
| **46** | 51 | 55 | 49 | 60 | 79 | 60 |
| **47** | 39 | 54 | 52 | 58 | 61 | 57 |
| **48** | 43 | 50 | 40 | 52 | 50 | 46 |
| **49** | 34 | 41 | 37 | 44 | 46 | 37 |
| **50** | 26 | 36 | 34 | 27 | 32 | 32 |

| Intensity | Port #1 | Port #2 | Port #3 | Port #4 | Port #5 | Port #b |
|---|---|---|---|---|---|---|
| **51** | 30 | 34 | 34 | 34 | 27 | 32 |
| **52** | 28 | 24 | 19 | 26 | 28 | 20 |
| **53** | 28 | 27 | 15 | 19 | 14 | 25 |
| **54** | 11 | 15 | 17 | 15 | 18 | 24 |
| **55** | 13 | 17 | 10 | 19 | 13 | 14 |
| **56** | 17 | 15 | 6 | 5 | 5 | 8 |
| **57** | 11 | 8 | 6 | 10 | 8 | 9 |
| **58** | 6 | 6 | 5 | 6 | 6 | 7 |
| **59** | 2 | 6 | 5 | 7 | 5 | 2 |
| **60** | 7 | 2 | 5 | 3 | 8 | 0 |
| **61** | 3 | 2 | 6 | 0 | 1 | 3 |
| **62** | 4 | 4 | 1 | 1 | 2 | 1 |
| **63** | 3 | 1 | 1 | 0 | 1 | 1 |
| **64** | 1 | 1 | 2 | 2 | 2 | 1 |
| **65** | 1 | 0 | 0 | 2 | 0 | 2 |
| **66** | 1 | 1 | 0 | 0 | 0 | 0 |
| **67** | 2 | 0 | 1 | 0 | 1 | 1 |
| **68** | 0 | 0 | 0 | 0 | 0 | 0 |
| **69** | 0 | 0 | 0 | 0 | 0 | 1 |
| **70** | 0 | 0 | 0 | 0 | 0 | 0 |
| **71** | 0 | 0 | 0 | 0 | 0 | 0 |
| **72** | 0 | 0 | 0 | 0 | 0 | 0 |
| **73** | 1 | 0 | 0 | 0 | 0 | 0 |
| **74** | 0 | 0 | 0 | 0 | 0 | 0 |
| **75** | 0 | 0 | 0 | 0 | 0 | 0 |
| **76** | 0 | 0 | 0 | 0 | 0 | 0 |
| **77** | 0 | 0 | 0 | 0 | 0 | 0 |
| **78** | 0 | 0 | 0 | 0 | 0 | 0 |
| **79** | 0 | 0 | 0 | 0 | 0 | 0 |
| **80** | 0 | 0 | 0 | 0 | 0 | 0 |
| **81** | 0 | 0 | 0 | 0 | 0 | 0 |
| **82** | 0 | 0 | 0 | 0 | 0 | 0 |
| **83** | 0 | 0 | 0 | 0 | 0 | 0 |
| **84** | 0 | 0 | 0 | 0 | 0 | 0 |
| **85** | 0 | 0 | 0 | 0 | 0 | 0 |
| **86** | 0 | 0 | 0 | 0 | 0 | 0 |
| **87** | 0 | 0 | 0 | 0 | 0 | 0 |
| **88** | 0 | 0 | 0 | 0 | 0 | 0 |
| **89** | 0 | 0 | 0 | 0 | 0 | 0 |
| **90** | 0 | 0 | 0 | 0 | 0 | 0 |
| **91** | 0 | 0 | 0 | 0 | 0 | 0 |
| **92** | 0 | 0 | 0 | 0 | 0 | 0 |
| **93** | 0 | 0 | 0 | 0 | 0 | 0 |
| **94** | 0 | 0 | 0 | 0 | 0 | 0 |
| **95** | 0 | 0 | 0 | 0 | 0 | 0 |
| **96** | 0 | 0 | 0 | 0 | 0 | 0 |
| **97** | 0 | 0 | 0 | 0 | 0 | 0 |
| **98** | 0 | 0 | 0 | 0 | 0 | 0 |
| **99** | 0 | 0 | 0 | 0 | 0 | 0 |
| **100** | 0 | 0 | 0 | 0 | 0 | 0 |

**Phase II Full-Scale Experiment**
**Test #** 4 **Chamber #** 4
**Pre-ozonated 0.94 $\mu$m YG-Microspheres**

| Intensity | Port #1 | Port #2 | Port #3 | Port #4 | Port #5 | Port #b |
|---|---|---|---|---|---|---|
| 1 | 15 | 12 | 7 | 3 | 17 | 3 |
| 2 | 76 | 78 | 72 | 47 | 78 | 56 |
| 3 | 169 | 126 | 119 | 90 | 131 | 105 |
| 4 | 129 | 122 | 93 | 97 | 110 | 103 |
| 5 | 96 | 76 | 62 | 60 | 61 | 71 |
| 6 | 51 | 42 | 30 | 31 | 36 | 35 |
| 7 | 33 | 20 | 29 | 20 | 29 | 12 |
| 8 | 13 | 14 | 13 | 3 | 17 | 10 |
| 9 | 11 | 18 | 9 | 9 | 15 | 16 |
| 10 | 12 | 8 | 10 | 11 | 8 | 12 |
| 11 | 12 | 9 | 16 | 10 | 19 | 4 |
| 12 | 8 | 17 | 10 | 5 | 13 | 0 |
| 13 | 11 | 8 | 13 | 9 | 11 | 7 |
| 14 | 16 | 10 | 23 | 18 | 13 | 15 |
| 15 | 26 | 21 | 15 | 25 | 9 | 17 |
| 16 | 25 | 31 | 19 | 18 | 22 | 24 |
| 17 | 36 | 37 | 23 | 32 | 26 | 26 |
| 18 | 35 | 37 | 42 | 34 | 38 | 28 |
| 19 | 37 | 47 | 39 | 43 | 43 | 33 |
| 20 | 50 | 43 | 62 | 56 | 51 | 33 |
| 21 | 46 | 61 | 58 | 53 | 56 | 41 |
| 22 | 65 | 79 | 60 | 78 | 71 | 55 |
| 23 | 75 | 92 | 96 | 94 | 88 | 42 |
| 24 | 81 | 98 | 91 | 87 | 102 | 57 |
| 25 | 104 | 96 | 86 | 106 | 100 | 52 |
| 26 | 110 | 110 | 98 | 99 | 101 | 78 |
| 27 | 113 | 132 | 102 | 139 | 102 | 66 |
| 28 | 138 | 120 | 113 | 134 | 118 | 60 |
| 29 | 144 | 128 | 125 | 108 | 143 | 78 |
| 30 | 114 | 148 | 134 | 153 | 145 | 73 |
| 31 | 113 | 129 | 118 | 135 | 143 | 91 |
| 32 | 147 | 128 | 107 | 148 | 127 | 76 |
| 33 | 148 | 155 | 119 | 136 | 109 | 77 |
| 34 | 144 | 114 | 120 | 119 | 100 | 102 |
| 35 | 112 | 136 | 114 | 119 | 130 | 71 |
| 36 | 102 | 125 | 137 | 113 | 131 | 83 |
| 37 | 109 | 104 | 114 | 92 | 115 | 72 |
| 38 | 101 | 120 | 103 | 117 | 90 | 71 |
| 39 | 103 | 114 | 94 | 105 | 103 | 64 |
| 40 | 93 | 99 | 97 | 84 | 78 | 61 |
| 41 | 96 | 84 | 99 | 93 | 86 | 49 |
| 42 | 72 | 93 | 83 | 68 | 77 | 62 |
| 43 | 75 | 86 | 68 | 79 | 77 | 54 |
| 44 | 70 | 73 | 63 | 68 | 78 | 44 |
| 45 | 61 | 76 | 58 | 61 | 57 | 24 |
| 46 | 67 | 55 | 63 | 54 | 69 | 46 |
| 47 | 49 | 49 | 50 | 45 | 56 | 30 |
| 48 | 52 | 38 | 30 | 47 | 29 | 34 |
| 49 | 35 | 31 | 35 | 32 | 32 | 22 |
| 50 | 32 | 39 | 30 | 26 | 39 | 18 |

| Intensity | Port #1 | Port #2 | Port #3 | Port #4 | Port #5 | Port #b |
|---|---|---|---|---|---|---|
| 51 | 39 | 33 | 29 | 34 | 24 | 15 |
| 52 | 25 | 20 | 28 | 22 | 25 | 16 |
| 53 | 16 | 22 | 22 | 15 | 26 | 15 |
| 54 | 18 | 17 | 22 | 13 | 16 | 9 |
| 55 | 15 | 23 | 14 | 9 | 10 | 9 |
| 56 | 12 | 10 | 7 | 9 | 7 | 11 |
| 57 | 14 | 18 | 8 | 5 | 10 | 4 |
| 58 | 3 | 8 | 6 | 8 | 6 | 5 |
| 59 | 8 | 3 | 9 | 4 | 3 | 4 |
| 60 | 0 | 3 | 3 | 3 | 1 | 0 |
| 61 | 0 | 3 | 2 | 0 | 5 | 2 |
| 62 | 0 | 1 | 0 | 1 | 2 | 1 |
| 63 | 2 | 1 | 1 | 0 | 0 | 2 |
| 64 | 0 | 1 | 1 | 0 | 0 | 2 |
| 65 | 0 | 0 | 1 | 1 | 0 | 0 |
| 66 | 0 | 1 | 1 | 2 | 0 | 1 |
| 67 | 0 | 0 | 0 | 0 | 0 | 0 |
| 68 | 0 | 0 | 0 | 1 | 0 | 0 |
| 69 | 0 | 0 | 0 | 0 | 0 | 0 |
| 70 | 1 | 0 | 0 | 0 | 0 | 0 |
| 71 | 0 | 0 | 0 | 0 | 0 | 0 |
| 72 | 0 | 0 | 0 | 0 | 0 | 0 |
| 73 | 0 | 0 | 0 | 0 | 0 | 0 |
| 74 | 0 | 0 | 0 | 0 | 0 | 0 |
| 75 | 0 | 0 | 0 | 0 | 0 | 0 |
| 76 | 0 | 0 | 0 | 0 | 0 | 0 |
| 77 | 0 | 0 | 0 | 0 | 0 | 1 |
| 78 | 0 | 0 | 0 | 0 | 0 | 0 |
| 79 | 0 | 0 | 0 | 0 | 0 | 0 |
| 80 | 0 | 0 | 0 | 0 | 0 | 0 |
| 81 | 0 | 0 | 0 | 0 | 0 | 0 |
| 82 | 0 | 0 | 0 | 0 | 0 | 0 |
| 83 | 0 | 0 | 0 | 0 | 0 | 0 |
| 84 | 0 | 0 | 0 | 0 | 0 | 0 |
| 85 | 0 | 0 | 0 | 0 | 0 | 0 |
| 86 | 0 | 0 | 0 | 0 | 0 | 0 |
| 87 | 0 | 0 | 0 | 0 | 0 | 0 |
| 88 | 0 | 0 | 0 | 0 | 0 | 0 |
| 89 | 0 | 0 | 0 | 0 | 0 | 0 |
| 90 | 0 | 0 | 0 | 0 | 0 | 0 |
| 91 | 0 | 0 | 0 | 0 | 0 | 0 |
| 92 | 0 | 0 | 0 | 0 | 0 | 0 |
| 93 | 0 | 0 | 0 | 0 | 0 | 0 |
| 94 | 0 | 0 | 0 | 0 | 0 | 0 |
| 95 | 0 | 0 | 0 | 0 | 0 | 0 |
| 96 | 0 | 0 | 0 | 0 | 0 | 0 |
| 97 | 0 | 0 | 0 | 0 | 0 | 0 |
| 98 | 0 | 0 | 0 | 0 | 0 | 0 |
| 99 | 0 | 0 | 0 | 0 | 0 | 0 |
| 100 | 0 | 0 | 0 | 0 | 0 | 0 |

**Phase II Full-Scale Experiment**
**Test #** 4 **Chamber #** 5
**Pre-ozonated 0.94 μm YG-Microspheres**

| Intensity | Port #1 | Port #2 | Port #3 | Port #4 | Port #5 | Port #b |
|---|---|---|---|---|---|---|
| **1** | 12 | | 6 | | 10 | 3 |
| **2** | 69 | | 71 | | 63 | 53 |
| **3** | 137 | | 152 | | 98 | 111 |
| **4** | 119 | | 115 | | 111 | 94 |
| **5** | 79 | | 76 | | 72 | 42 |
| **6** | 43 | | 41 | | 42 | 28 |
| **7** | 20 | | 14 | | 23 | 18 |
| **8** | 8 | | 13 | | 13 | 14 |
| **9** | 9 | | 10 | | 7 | 7 |
| **10** | 7 | | 8 | | 9 | 7 |
| **11** | 14 | | 6 | | 14 | 9 |
| **12** | 15 | | 10 | | 12 | 7 |
| **13** | 11 | | 9 | | 8 | 15 |
| **14** | 17 | | 10 | | 16 | 16 |
| **15** | 13 | | 13 | | 8 | 20 |
| **16** | 18 | | 27 | | 23 | 28 |
| **17** | 27 | | 22 | | 17 | 32 |
| **18** | 39 | | 27 | | 34 | 40 |
| **19** | 50 | | 42 | | 47 | 47 |
| **20** | 63 | | 43 | | 32 | 53 |
| **21** | 76 | | 60 | | 67 | 80 |
| **22** | 74 | | 47 | | 71 | 76 |
| **23** | 75 | | 64 | | 76 | 76 |
| **24** | 96 | | 77 | | 93 | 81 |
| **25** | 89 | | 88 | | 84 | 89 |
| **26** | 123 | | 75 | | 112 | 102 |
| **27** | 126 | | 107 | | 130 | 110 |
| **28** | 116 | | 102 | | 130 | 110 |
| **29** | 170 | | 108 | | 114 | 124 |
| **30** | 135 | | 119 | | 143 | 129 |
| **31** | 142 | | 95 | | 140 | 133 |
| **32** | 101 | | 91 | | 132 | 122 |
| **33** | 157 | | 96 | | 122 | 117 |
| **34** | 140 | | 119 | | 127 | 125 |
| **35** | 115 | | 116 | | 116 | 122 |
| **36** | 141 | | 97 | | 115 | 105 |
| **37** | 121 | | 91 | | 108 | 123 |
| **38** | 112 | | 92 | | 103 | 86 |
| **39** | 94 | | 81 | | 110 | 98 |
| **40** | 88 | | 82 | | 96 | 84 |
| **41** | 102 | | 75 | | 89 | 91 |
| **42** | 67 | | 68 | | 70 | 86 |
| **43** | 54 | | 51 | | 89 | 65 |
| **44** | 84 | | 55 | | 69 | 71 |
| **45** | 72 | | 62 | | 56 | 58 |
| **46** | 53 | | 47 | | 56 | 59 |
| **47** | 55 | | 30 | | 77 | 37 |
| **48** | 45 | | 39 | | 47 | 43 |
| **49** | 46 | | 27 | | 47 | 47 |
| **50** | 31 | | 27 | | 27 | 39 |

| Intensity | Port #1 | Port #2 | Port #3 | Port #4 | Port #5 | Port #b |
|---|---|---|---|---|---|---|
| **51** | 27 | | 23 | | 26 | 27 |
| **52** | 23 | | 23 | | 18 | 23 |
| **53** | 19 | | 13 | | 23 | 12 |
| **54** | 12 | | 12 | | 17 | 12 |
| **55** | 18 | | 3 | | 12 | 8 |
| **56** | 15 | | 4 | | 10 | 19 |
| **57** | 7 | | 6 | | 9 | 4 |
| **58** | 6 | | 2 | | 5 | 3 |
| **59** | 4 | | 5 | | 3 | 1 |
| **60** | 4 | | 1 | | 2 | 0 |
| **61** | 1 | | 1 | | 1 | 0 |
| **62** | 0 | | 1 | | 3 | 1 |
| **63** | 2 | | 0 | | 1 | 1 |
| **64** | 0 | | 0 | | 3 | 1 |
| **65** | 1 | | 0 | | 3 | 0 |
| **66** | 0 | | 0 | | 0 | 0 |
| **67** | 0 | | 0 | | 0 | 0 |
| **68** | 0 | | 0 | | 0 | 0 |
| **69** | 0 | | 0 | | 0 | 0 |
| **70** | 0 | | 0 | | 0 | 0 |
| **71** | 0 | | 0 | | 0 | 0 |
| **72** | 0 | | 1 | | 0 | 0 |
| **73** | 0 | | 0 | | 0 | 0 |
| **74** | 0 | | 0 | | 0 | 0 |
| **75** | 0 | | 0 | | 0 | 0 |
| **76** | 0 | | 0 | | 0 | 0 |
| **77** | 0 | | 0 | | 0 | 0 |
| **78** | 0 | | 0 | | 0 | 0 |
| **79** | 0 | | 0 | | 0 | 0 |
| **80** | 0 | | 0 | | 0 | 0 |
| **81** | 0 | | 0 | | 0 | 0 |
| **82** | 0 | | 0 | | 0 | 0 |
| **83** | 0 | | 0 | | 0 | 0 |
| **84** | 0 | | 0 | | 0 | 0 |
| **85** | 0 | | 0 | | 0 | 0 |
| **86** | 0 | | 0 | | 0 | 0 |
| **87** | 0 | | 0 | | 0 | 0 |
| **88** | 0 | | 0 | | 0 | 0 |
| **89** | 0 | | 0 | | 0 | 0 |
| **90** | 0 | | 0 | | 0 | 0 |
| **91** | 0 | | 0 | | 0 | 0 |
| **92** | 0 | | 0 | | 0 | 0 |
| **93** | 0 | | 0 | | 0 | 0 |
| **94** | 0 | | 0 | | 0 | 0 |
| **95** | 0 | | 0 | | 0 | 0 |
| **96** | 0 | | 0 | | 0 | 0 |
| **97** | 0 | | 0 | | 0 | 0 |
| **98** | 0 | | 0 | | 0 | 0 |
| **99** | 0 | | 0 | | 0 | 0 |
| **100** | 0 | | 0 | | 0 | 0 |

**Phase II Full-Scale Experiment**
**Test #** 5 **Chamber #** 1
**Pre-ozonated 0.94 $\mu$m YG-Microspheres**

| Intensity | Port #1 | Port #2 | Port #3 | Port #4 | Port #5 | Port #b |
|---|---|---|---|---|---|---|
| 1 | 5 | 13 | 12 | 11 | 8 | 12 |
| 2 | 96 | 111 | 89 | 71 | 58 | 42 |
| 3 | 152 | 213 | 167 | 141 | 137 | 91 |
| 4 | 163 | 244 | 162 | 122 | 131 | 106 |
| 5 | 153 | 153 | 104 | 93 | 70 | 68 |
| 6 | 97 | 111 | 76 | 42 | 51 | 43 |
| 7 | 50 | 53 | 48 | 36 | 31 | 22 |
| 8 | 32 | 34 | 22 | 18 | 15 | 12 |
| 9 | 21 | 16 | 18 | 18 | 12 | 4 |
| 10 | 10 | 16 | 12 | 12 | 6 | 10 |
| 11 | 7 | 11 | 9 | 7 | 7 | 2 |
| 12 | 6 | 11 | 10 | 4 | 9 | 7 |
| 13 | 10 | 9 | 7 | 4 | 7 | 5 |
| 14 | 4 | 6 | 8 | 10 | 6 | 5 |
| 15 | 9 | 5 | 11 | 9 | 7 | 9 |
| 16 | 10 | 6 | 14 | 9 | 7 | 15 |
| 17 | 14 | 7 | 13 | 17 | 11 | 24 |
| 18 | 14 | 15 | 25 | 10 | 13 | 15 |
| 19 | 15 | 10 | 20 | 24 | 24 | 34 |
| 20 | 20 | 21 | 28 | 14 | 19 | 28 |
| 21 | 25 | 17 | 32 | 27 | 18 | 44 |
| 22 | 38 | 25 | 34 | 35 | 26 | 43 |
| 23 | 38 | 31 | 35 | 28 | 22 | 51 |
| 24 | 36 | 33 | 54 | 42 | 39 | 59 |
| 25 | 54 | 57 | 62 | 45 | 35 | 83 |
| 26 | 62 | 41 | 69 | 46 | 59 | 72 |
| 27 | 51 | 57 | 91 | 56 | 57 | 85 |
| 28 | 58 | 74 | 73 | 79 | 69 | 89 |
| 29 | 76 | 80 | 77 | 67 | 69 | 77 |
| 30 | 74 | 70 | 78 | 79 | 89 | 102 |
| 31 | 82 | 94 | 93 | 92 | 73 | 124 |
| 32 | 78 | 64 | 96 | 94 | 80 | 109 |
| 33 | 98 | 88 | 104 | 82 | 66 | 107 |
| 34 | 106 | 86 | 95 | 91 | 97 | 103 |
| 35 | 91 | 91 | 108 | 78 | 93 | 124 |
| 36 | 99 | 87 | 105 | 95 | 114 | 116 |
| 37 | 96 | 106 | 117 | 98 | 101 | 121 |
| 38 | 99 | 104 | 114 | 118 | 99 | 128 |
| 39 | 99 | 110 | 119 | 112 | 85 | 122 |
| 40 | 91 | 119 | 113 | 109 | 111 | 101 |
| 41 | 107 | 89 | 113 | 100 | 98 | 115 |
| 42 | 106 | 99 | 95 | 88 | 84 | 95 |
| 43 | 101 | 92 | 96 | 103 | 108 | 82 |
| 44 | 88 | 94 | 90 | 77 | 94 | 90 |
| 45 | 85 | 98 | 82 | 72 | 98 | 83 |
| 46 | 78 | 89 | 87 | 83 | 95 | 78 |
| 47 | 91 | 94 | 74 | 100 | 101 | 83 |
| 48 | 71 | 84 | 73 | 69 | 76 | 73 |
| 49 | 81 | 68 | 86 | 84 | 69 | 64 |
| 50 | 84 | 80 | 61 | 80 | 85 | 57 |

| Intensity | Port #1 | Port #2 | Port #3 | Port #4 | Port #5 | Port #b |
|---|---|---|---|---|---|---|
| 51 | 58 | 58 | 58 | 63 | 65 | 53 |
| 52 | 40 | 56 | 48 | 63 | 64 | 50 |
| 53 | 52 | 73 | 39 | 54 | 53 | 43 |
| 54 | 58 | 59 | 41 | 53 | 58 | 27 |
| 55 | 48 | 45 | 40 | 59 | 56 | 34 |
| 56 | 57 | 64 | 39 | 47 | 40 | 18 |
| 57 | 49 | 42 | 19 | 37 | 39 | 29 |
| 58 | 38 | 37 | 19 | 42 | 40 | 21 |
| 59 | 32 | 42 | 25 | 27 | 28 | 22 |
| 60 | 38 | 31 | 10 | 26 | 32 | 22 |
| 61 | 24 | 25 | 12 | 17 | 31 | 11 |
| 62 | 28 | 25 | 5 | 30 | 24 | 3 |
| 63 | 22 | 27 | 4 | 20 | 20 | 5 |
| 64 | 17 | 11 | 5 | 11 | 18 | 11 |
| 65 | 16 | 15 | 5 | 14 | 22 | 4 |
| 66 | 22 | 20 | 3 | 9 | 18 | 6 |
| 67 | 13 | 12 | 3 | 9 | 17 | 5 |
| 68 | 10 | 9 | 0 | 7 | 12 | 0 |
| 69 | 13 | 5 | 1 | 8 | 10 | 2 |
| 70 | 9 | 7 | 2 | 6 | 11 | 1 |
| 71 | 3 | 6 | 3 | 3 | 8 | 0 |
| 72 | 12 | 3 | 1 | 3 | 6 | 1 |
| 73 | 4 | 1 | 1 | 2 | 5 | 0 |
| 74 | 4 | 5 | 0 | 2 | 5 | 0 |
| 75 | 3 | 5 | 1 | 0 | 2 | 1 |
| 76 | 1 | 2 | 2 | 1 | 2 | 0 |
| 77 | 2 | 4 | 0 | 1 | 1 | 0 |
| 78 | 1 | 1 | 0 | 2 | 1 | 1 |
| 79 | 1 | 0 | 1 | 0 | 0 | 0 |
| 80 | 1 | 1 | 0 | 0 | 0 | 0 |
| 81 | 2 | 0 | 0 | 0 | 0 | 0 |
| 82 | 0 | 0 | 0 | 1 | 1 | 0 |
| 83 | 1 | 0 | 0 | 0 | 0 | 0 |
| 84 | 0 | 2 | 0 | 0 | 0 | 0 |
| 85 | 0 | 0 | 0 | 0 | 0 | 0 |
| 86 | 2 | 0 | 0 | 0 | 0 | 0 |
| 87 | 0 | 0 | 0 | 0 | 0 | 0 |
| 88 | 0 | 0 | 0 | 0 | 0 | 0 |
| 89 | 0 | 0 | 0 | 0 | 0 | 0 |
| 90 | 0 | 0 | 0 | 0 | 0 | 0 |
| 91 | 0 | 0 | 0 | 0 | 0 | 0 |
| 92 | 0 | 0 | 0 | 0 | 0 | 0 |
| 93 | 0 | 0 | 0 | 0 | 0 | 0 |
| 94 | 0 | 0 | 0 | 0 | 0 | 1 |
| 95 | 0 | 0 | 0 | 0 | 0 | 0 |
| 96 | 0 | 0 | 0 | 0 | 0 | 0 |
| 97 | 0 | 0 | 0 | 0 | 0 | 0 |
| 98 | 0 | 0 | 0 | 0 | 0 | 0 |
| 99 | 0 | 0 | 0 | 0 | 0 | 0 |
| 100 | 0 | 0 | 0 | 0 | 0 | 0 |

**Phase II Full-Scale Experiment**
**Test #** 5 **Chamber #** 2
**Pre-ozonated 0.94 $\mu$m YG-Microspheres**

| Intensity | Port #1 | Port #2 | Port #3 | Port #4 | Port #5 | Port #b |
|---|---|---|---|---|---|---|
| 1 | 2 | 8 | 15 | 9 | 4 | |
| 2 | 67 | 63 | 56 | 70 | 67 | |
| 3 | 143 | 155 | 151 | 137 | 108 | |
| 4 | 141 | 161 | 125 | 146 | 121 | |
| 5 | 88 | 99 | 94 | 90 | 87 | |
| 6 | 54 | 56 | 60 | 54 | 62 | |
| 7 | 28 | 43 | 34 | 28 | 27 | |
| 8 | 23 | 21 | 15 | 22 | 20 | |
| 9 | 13 | 11 | 20 | 6 | 9 | |
| 10 | 5 | 12 | 9 | 9 | 11 | |
| 11 | 9 | 9 | 12 | 11 | 5 | |
| 12 | 4 | 6 | 13 | 8 | 8 | |
| 13 | 11 | 11 | 15 | 11 | 11 | |
| 14 | 7 | 12 | 11 | 10 | 5 | |
| 15 | 11 | 11 | 18 | 11 | 12 | |
| 16 | 14 | 21 | 15 | 15 | 21 | |
| 17 | 21 | 29 | 15 | 12 | 22 | |
| 18 | 23 | 28 | 15 | 17 | 34 | |
| 19 | 35 | 31 | 34 | 32 | 19 | |
| 20 | 31 | 26 | 35 | 43 | 48 | |
| 21 | 46 | 46 | 35 | 37 | 34 | |
| 22 | 51 | 66 | 54 | 39 | 45 | |
| 23 | 73 | 56 | 59 | 44 | 61 | |
| 24 | 50 | 54 | 64 | 59 | 71 | |
| 25 | 93 | 71 | 89 | 81 | 68 | |
| 26 | 83 | 76 | 93 | 77 | 83 | |
| 27 | 91 | 96 | 96 | 78 | 97 | |
| 28 | 103 | 77 | 103 | 101 | 92 | |
| 29 | 103 | 87 | 101 | 97 | 116 | |
| 30 | 109 | 103 | 123 | 96 | 130 | |
| 31 | 114 | 111 | 124 | 99 | 113 | |
| 32 | 150 | 144 | 121 | 106 | 126 | |
| 33 | 113 | 103 | 128 | 120 | 129 | |
| 34 | 115 | 102 | 114 | 100 | 125 | |
| 35 | 106 | 91 | 107 | 132 | 118 | |
| 36 | 124 | 103 | 103 | 99 | 131 | |
| 37 | 133 | 123 | 139 | 118 | 133 | |
| 38 | 110 | 131 | 115 | 121 | 120 | |
| 39 | 93 | 135 | 116 | 87 | 114 | |
| 40 | 112 | 106 | 99 | 102 | 97 | |
| 41 | 103 | 112 | 86 | 97 | 94 | |
| 42 | 98 | 88 | 85 | 95 | 93 | |
| 43 | 95 | 88 | 90 | 99 | 91 | |
| 44 | 96 | 96 | 83 | 79 | 81 | |
| 45 | 66 | 77 | 71 | 80 | 72 | |
| 46 | 84 | 85 | 59 | 86 | 59 | |
| 47 | 52 | 69 | 57 | 64 | 60 | |
| 48 | 61 | 69 | 62 | 66 | 50 | |
| 49 | 51 | 53 | 47 | 64 | 54 | |
| 50 | 53 | 53 | 46 | 53 | 52 | |

| Intensity | Port #1 | Port #2 | Port #3 | Port #4 | Port #5 | Port #b |
|---|---|---|---|---|---|---|
| 51 | 43 | 39 | 35 | 54 | 38 | |
| 52 | 43 | 27 | 40 | 52 | 28 | |
| 53 | 30 | 27 | 25 | 40 | 36 | |
| 54 | 37 | 29 | 19 | 33 | 33 | |
| 55 | 28 | 25 | 19 | 36 | 24 | |
| 56 | 18 | 19 | 15 | 18 | 24 | |
| 57 | 17 | 18 | 22 | 27 | 15 | |
| 58 | 8 | 20 | 11 | 20 | 23 | |
| 59 | 11 | 10 | 11 | 16 | 7 | |
| 60 | 7 | 9 | 8 | 7 | 7 | |
| 61 | 4 | 11 | 12 | 8 | 6 | |
| 62 | 3 | 3 | 8 | 4 | 5 | |
| 63 | 6 | 11 | 5 | 7 | 8 | |
| 64 | 1 | 6 | 5 | 6 | 2 | |
| 65 | 3 | 2 | 4 | 5 | 3 | |
| 66 | 1 | 1 | 3 | 5 | 0 | |
| 67 | 1 | 1 | 2 | 0 | 3 | |
| 68 | 0 | 0 | 1 | 1 | 0 | |
| 69 | 2 | 1 | 0 | 1 | 2 | |
| 70 | 2 | 1 | 1 | 0 | 1 | |
| 71 | 0 | 0 | 0 | 0 | 1 | |
| 72 | 0 | 1 | 0 | 1 | 0 | |
| 73 | 0 | 0 | 0 | 0 | 0 | |
| 74 | 0 | 0 | 0 | 2 | 0 | |
| 75 | 2 | 0 | 0 | 0 | 0 | |
| 76 | 0 | 0 | 0 | 0 | 0 | |
| 77 | 0 | 0 | 0 | 1 | 0 | |
| 78 | 0 | 1 | 0 | 0 | 0 | |
| 79 | 0 | 0 | 0 | 0 | 0 | |
| 80 | 0 | 0 | 0 | 0 | 0 | |
| 81 | 0 | 0 | 0 | 0 | 0 | |
| 82 | 0 | 0 | 0 | 0 | 0 | |
| 83 | 0 | 0 | 0 | 0 | 0 | |
| 84 | 0 | 0 | 0 | 0 | 0 | |
| 85 | 0 | 0 | 0 | 0 | 0 | |
| 86 | 0 | 0 | 0 | 0 | 0 | |
| 87 | 0 | 1 | 0 | 0 | 0 | |
| 88 | 0 | 0 | 0 | 0 | 0 | |
| 89 | 0 | 0 | 0 | 0 | 0 | |
| 90 | 0 | 0 | 0 | 0 | 0 | |
| 91 | 0 | 0 | 0 | 0 | 0 | |
| 92 | 0 | 0 | 0 | 0 | 0 | |
| 93 | 0 | 0 | 0 | 0 | 0 | |
| 94 | 0 | 0 | 0 | 0 | 0 | |
| 95 | 0 | 0 | 0 | 0 | 0 | |
| 96 | 0 | 0 | 0 | 0 | 0 | |
| 97 | 0 | 0 | 0 | 0 | 0 | |
| 98 | 0 | 0 | 0 | 0 | 0 | |
| 99 | 0 | 0 | 0 | 0 | 0 | |
| 100 | 0 | 0 | 0 | 0 | 0 | |

**Phase II Full-Scale Experiment**
**Test #** 5 **Chamber #** 3
**Pre-ozonated 0.94 μm YG-Microspheres**

| Intensity | Port #1 | Port #2 | Port #3 | Port #4 | Port #5 | Port #b |
|---|---|---|---|---|---|---|
| 1 | 10 | 11 | 9 | 10 | 7 | 7 |
| 2 | 75 | 79 | 95 | 68 | 79 | 61 |
| 3 | 152 | 151 | 180 | 173 | 170 | 142 |
| 4 | 141 | 146 | 177 | 163 | 163 | 115 |
| 5 | 102 | 86 | 113 | 107 | 127 | 82 |
| 6 | 76 | 61 | 83 | 52 | 54 | 35 |
| 7 | 34 | 34 | 46 | 24 | 33 | 25 |
| 8 | 22 | 16 | 27 | 22 | 30 | 19 |
| 9 | 9 | 13 | 15 | 14 | 12 | 5 |
| 10 | 14 | 10 | 12 | 17 | 11 | 5 |
| 11 | 11 | 11 | 7 | 18 | 14 | 6 |
| 12 | 12 | 9 | 14 | 4 | 14 | 7 |
| 13 | 10 | 7 | 12 | 6 | 10 | 12 |
| 14 | 7 | 11 | 9 | 16 | 15 | 12 |
| 15 | 15 | 16 | 13 | 10 | 12 | 15 |
| 16 | 19 | 10 | 10 | 9 | 16 | 12 |
| 17 | 18 | 15 | 19 | 17 | 21 | 16 |
| 18 | 20 | 28 | 17 | 22 | 26 | 22 |
| 19 | 25 | 25 | 28 | 24 | 38 | 33 |
| 20 | 40 | 24 | 40 | 36 | 39 | 40 |
| 21 | 44 | 46 | 42 | 36 | 39 | 32 |
| 22 | 53 | 39 | 41 | 41 | 45 | 49 |
| 23 | 58 | 53 | 53 | 57 | 62 | 51 |
| 24 | 72 | 79 | 65 | 78 | 65 | 67 |
| 25 | 77 | 73 | 76 | 64 | 73 | 91 |
| 26 | 88 | 81 | 86 | 88 | 103 | 83 |
| 27 | 106 | 105 | 81 | 79 | 89 | 93 |
| 28 | 93 | 87 | 93 | 90 | 97 | 94 |
| 29 | 85 | 103 | 114 | 112 | 119 | 102 |
| 30 | 100 | 109 | 107 | 108 | 108 | 108 |
| 31 | 121 | 136 | 129 | 125 | 117 | 103 |
| 32 | 119 | 115 | 120 | 115 | 132 | 106 |
| 33 | 106 | 132 | 112 | 128 | 131 | 120 |
| 34 | 114 | 126 | 112 | 99 | 124 | 97 |
| 35 | 107 | 123 | 103 | 120 | 118 | 125 |
| 36 | 114 | 114 | 114 | 104 | 121 | 95 |
| 37 | 121 | 122 | 114 | 92 | 130 | 127 |
| 38 | 117 | 111 | 105 | 108 | 124 | 113 |
| 39 | 117 | 108 | 104 | 103 | 109 | 104 |
| 40 | 103 | 94 | 115 | 98 | 99 | 105 |
| 41 | 106 | 117 | 91 | 108 | 106 | 96 |
| 42 | 84 | 91 | 79 | 95 | 88 | 104 |
| 43 | 97 | 85 | 96 | 111 | 92 | 87 |
| 44 | 81 | 69 | 95 | 84 | 72 | 81 |
| 45 | 90 | 87 | 79 | 83 | 70 | 82 |
| 46 | 68 | 73 | 69 | 78 | 89 | 76 |
| 47 | 76 | 57 | 63 | 75 | 57 | 77 |
| 48 | 73 | 59 | 72 | 53 | 60 | 69 |
| 49 | 52 | 49 | 48 | 44 | 52 | 64 |
| 50 | 49 | 47 | 49 | 49 | 47 | 56 |

| Intensity | Port #1 | Port #2 | Port #3 | Port #4 | Port #5 | Port #b |
|---|---|---|---|---|---|---|
| 51 | 57 | 46 | 38 | 53 | 37 | 44 |
| 52 | 39 | 32 | 30 | 28 | 41 | 39 |
| 53 | 32 | 31 | 36 | 31 | 52 | 25 |
| 54 | 23 | 27 | 30 | 23 | 32 | 27 |
| 55 | 22 | 21 | 24 | 31 | 18 | 20 |
| 56 | 20 | 20 | 26 | 16 | 16 | 17 |
| 57 | 18 | 19 | 16 | 12 | 12 | 21 |
| 58 | 20 | 14 | 20 | 16 | 11 | 11 |
| 59 | 17 | 11 | 17 | 8 | 10 | 9 |
| 60 | 5 | 6 | 5 | 4 | 8 | 5 |
| 61 | 4 | 8 | 9 | 3 | 5 | 3 |
| 62 | 1 | 2 | 4 | 7 | 3 | 1 |
| 63 | 1 | 6 | 8 | 3 | 3 | 4 |
| 64 | 0 | 2 | 3 | 3 | 2 | 2 |
| 65 | 3 | 5 | 3 | 4 | 3 | 2 |
| 66 | 0 | 1 | 3 | 1 | 3 | 1 |
| 67 | 0 | 0 | 6 | 0 | 1 | 2 |
| 68 | 1 | 1 | 2 | 0 | 0 | 1 |
| 69 | 2 | 0 | 0 | 0 | 0 | 1 |
| 70 | 0 | 0 | 0 | 3 | 0 | 1 |
| 71 | 0 | 0 | 0 | 0 | 0 | 1 |
| 72 | 0 | 0 | 0 | 1 | 0 | 0 |
| 73 | 0 | 0 | 0 | 0 | 0 | 0 |
| 74 | 0 | 0 | 0 | 0 | 0 | 0 |
| 75 | 0 | 0 | 0 | 0 | 0 | 0 |
| 76 | 0 | 0 | 0 | 0 | 0 | 1 |
| 77 | 0 | 0 | 0 | 0 | 1 | 1 |
| 78 | 0 | 0 | 0 | 0 | 1 | 0 |
| 79 | 0 | 0 | 0 | 0 | 0 | 0 |
| 80 | 0 | 0 | 0 | 0 | 0 | 0 |
| 81 | 0 | 0 | 0 | 0 | 0 | 0 |
| 82 | 0 | 0 | 0 | 0 | 0 | 0 |
| 83 | 0 | 0 | 0 | 0 | 0 | 0 |
| 84 | 0 | 0 | 0 | 0 | 0 | 0 |
| 85 | 0 | 0 | 0 | 0 | 0 | 0 |
| 86 | 0 | 0 | 0 | 0 | 0 | 0 |
| 87 | 0 | 1 | 0 | 0 | 0 | 0 |
| 88 | 0 | 0 | 0 | 0 | 0 | 0 |
| 89 | 0 | 0 | 0 | 0 | 0 | 0 |
| 90 | 0 | 0 | 0 | 0 | 0 | 0 |
| 91 | 0 | 0 | 0 | 0 | 0 | 0 |
| 92 | 0 | 0 | 0 | 0 | 0 | 0 |
| 93 | 0 | 0 | 0 | 0 | 0 | 0 |
| 94 | 0 | 0 | 0 | 0 | 0 | 0 |
| 95 | 0 | 0 | 0 | 0 | 0 | 0 |
| 96 | 0 | 0 | 0 | 0 | 0 | 0 |
| 97 | 0 | 0 | 0 | 0 | 0 | 0 |
| 98 | 0 | 0 | 0 | 0 | 0 | 0 |
| 99 | 0 | 0 | 0 | 0 | 0 | 0 |
| 100 | 0 | 0 | 0 | 0 | 0 | 0 |

**Phase II Full-Scale Experiment**
**Test #** 5 **Chamber #** 4
**Pre-ozonated 0.94 μm YG-Microspheres**

| Intensity | Port #1 | Port #2 | Port #3 | Port #4 | Port #5 | Port #b |
|---|---|---|---|---|---|---|
| **1** | 7 | | 3 | | 14 | 5 |
| **2** | 59 | | 40 | | 65 | 59 |
| **3** | 137 | | 82 | | 136 | 144 |
| **4** | 117 | | 78 | | 124 | 110 |
| **5** | 58 | | 35 | | 74 | 73 |
| **6** | 45 | | 20 | | 43 | 38 |
| **7** | 19 | | 8 | | 21 | 21 |
| **8** | 16 | | 9 | | 9 | 12 |
| **9** | 6 | | 6 | | 6 | 5 |
| **10** | 8 | | 2 | | 1 | 7 |
| **11** | 3 | | 4 | | 8 | 7 |
| **12** | 11 | | 8 | | 6 | 6 |
| **13** | 8 | | 5 | | 5 | 6 |
| **14** | 9 | | 9 | | 7 | 14 |
| **15** | 10 | | 12 | | 20 | 12 |
| **16** | 18 | | 21 | | 23 | 12 |
| **17** | 19 | | 12 | | 29 | 20 |
| **18** | 22 | | 18 | | 19 | 20 |
| **19** | 28 | | 29 | | 28 | 37 |
| **20** | 44 | | 30 | | 40 | 28 |
| **21** | 39 | | 43 | | 49 | 34 |
| **22** | 54 | | 64 | | 44 | 47 |
| **23** | 59 | | 42 | | 56 | 63 |
| **24** | 65 | | 57 | | 84 | 71 |
| **25** | 86 | | 72 | | 94 | 90 |
| **26** | 108 | | 80 | | 89 | 86 |
| **27** | 84 | | 91 | | 90 | 98 |
| **28** | 116 | | 97 | | 99 | 100 |
| **29** | 106 | | 96 | | 99 | 126 |
| **30** | 110 | | 112 | | 109 | 104 |
| **31** | 113 | | 93 | | 110 | 135 |
| **32** | 109 | | 126 | | 116 | 123 |
| **33** | 101 | | 140 | | 112 | 136 |
| **34** | 114 | | 137 | | 122 | 127 |
| **35** | 142 | | 84 | | 118 | 126 |
| **36** | 109 | | 113 | | 115 | 108 |
| **37** | 112 | | 125 | | 106 | 103 |
| **38** | 114 | | 114 | | 130 | 121 |
| **39** | 112 | | 95 | | 120 | 106 |
| **40** | 108 | | 110 | | 111 | 109 |
| **41** | 104 | | 106 | | 95 | 100 |
| **42** | 84 | | 112 | | 96 | 81 |
| **43** | 94 | | 95 | | 109 | 79 |
| **44** | 95 | | 86 | | 85 | 90 |
| **45** | 66 | | 81 | | 69 | 71 |
| **46** | 78 | | 78 | | 74 | 71 |
| **47** | 63 | | 60 | | 66 | 50 |
| **48** | 65 | | 58 | | 50 | 57 |
| **49** | 55 | | 45 | | 58 | 54 |
| **50** | 47 | | 35 | | 44 | 41 |

| Intensity | Port #1 | Port #2 | Port #3 | Port #4 | Port #5 | Port #b |
|---|---|---|---|---|---|---|
| **51** | 42 | | 41 | | 44 | 39 |
| **52** | 31 | | 42 | | 30 | 30 |
| **53** | 30 | | 37 | | 24 | 33 |
| **54** | 22 | | 23 | | 31 | 26 |
| **55** | 22 | | 22 | | 10 | 18 |
| **56** | 15 | | 19 | | 3 | 19 |
| **57** | 18 | | 19 | | 14 | 13 |
| **58** | 11 | | 17 | | 13 | 11 |
| **59** | 5 | | 14 | | 7 | 11 |
| **60** | 5 | | 12 | | 9 | 12 |
| **61** | 3 | | 0 | | 2 | 2 |
| **62** | 2 | | 7 | | 5 | 1 |
| **63** | 4 | | 0 | | 7 | 2 |
| **64** | 2 | | 2 | | 2 | 0 |
| **65** | 0 | | 1 | | 1 | 0 |
| **66** | 2 | | 3 | | 2 | 0 |
| **67** | 1 | | 1 | | 1 | 0 |
| **68** | 0 | | 1 | | 1 | 3 |
| **69** | 0 | | 0 | | 0 | 2 |
| **70** | 0 | | 2 | | 0 | 1 |
| **71** | 0 | | 0 | | 0 | 1 |
| **72** | 1 | | 0 | | 0 | 0 |
| **73** | 0 | | 0 | | 0 | 0 |
| **74** | 0 | | 0 | | 0 | 0 |
| **75** | 0 | | 0 | | 0 | 0 |
| **76** | 0 | | 0 | | 0 | 0 |
| **77** | 0 | | 1 | | 0 | 0 |
| **78** | 0 | | 0 | | 0 | 0 |
| **79** | 0 | | 0 | | 0 | 0 |
| **80** | 0 | | 0 | | 0 | 0 |
| **81** | 0 | | 0 | | 0 | 0 |
| **82** | 0 | | 0 | | 0 | 0 |
| **83** | 0 | | 0 | | 0 | 0 |
| **84** | 0 | | 0 | | 0 | 0 |
| **85** | 0 | | 0 | | 0 | 0 |
| **86** | 0 | | 0 | | 0 | 0 |
| **87** | 0 | | 0 | | 0 | 0 |
| **88** | 0 | | 0 | | 0 | 0 |
| **89** | 0 | | 0 | | 0 | 0 |
| **90** | 0 | | 0 | | 0 | 0 |
| **91** | 0 | | 0 | | 0 | 0 |
| **92** | 0 | | 0 | | 0 | 0 |
| **93** | 0 | | 0 | | 0 | 0 |
| **94** | 0 | | 0 | | 0 | 0 |
| **95** | 0 | | 0 | | 0 | 0 |
| **96** | 0 | | 0 | | 0 | 0 |
| **97** | 0 | | 0 | | 0 | 0 |
| **98** | 0 | | 0 | | 0 | 0 |
| **99** | 0 | | 0 | | 0 | 0 |
| **100** | 0 | | 0 | | 0 | 0 |

**Phase II Full-Scale Experiment**
**Test #** 5 **Chamber #** 5
**Pre-ozonated 0.94 μm YG-Microspheres**

| Intensity | Port #1 | Port #2 | Port #3 | Port #4 | Port #5 | Port #b |
|---|---|---|---|---|---|---|
| **1** | 6 | | 6 | | 10 | 7 |
| **2** | 47 | | 54 | | 68 | 59 |
| **3** | 70 | | 114 | | 123 | 142 |
| **4** | 77 | | 89 | | 116 | 111 |
| **5** | 47 | | 68 | | 73 | 66 |
| **6** | 30 | | 33 | | 45 | 32 |
| **7** | 15 | | 16 | | 26 | 23 |
| **8** | 11 | | 12 | | 14 | 16 |
| **9** | 4 | | 8 | | 13 | 3 |
| **10** | 7 | | 4 | | 4 | 5 |
| **11** | 4 | | 7 | | 10 | 6 |
| **12** | 9 | | 7 | | 3 | 10 |
| **13** | 10 | | 7 | | 5 | 6 |
| **14** | 6 | | 7 | | 6 | 8 |
| **15** | 24 | | 12 | | 13 | 16 |
| **16** | 19 | | 14 | | 14 | 17 |
| **17** | 19 | | 18 | | 24 | 17 |
| **18** | 25 | | 30 | | 26 | 24 |
| **19** | 43 | | 21 | | 33 | 29 |
| **20** | 30 | | 34 | | 32 | 41 |
| **21** | 43 | | 46 | | 49 | 51 |
| **22** | 44 | | 49 | | 55 | 49 |
| **23** | 53 | | 52 | | 72 | 60 |
| **24** | 65 | | 70 | | 67 | 60 |
| **25** | 76 | | 83 | | 66 | 61 |
| **26** | 77 | | 74 | | 91 | 83 |
| **27** | 93 | | 96 | | 104 | 87 |
| **28** | 92 | | 79 | | 102 | 98 |
| **29** | 96 | | 111 | | 117 | 94 |
| **30** | 100 | | 91 | | 103 | 106 |
| **31** | 120 | | 116 | | 112 | 118 |
| **32** | 117 | | 124 | | 104 | 130 |
| **33** | 129 | | 122 | | 111 | 124 |
| **34** | 117 | | 136 | | 127 | 133 |
| **35** | 122 | | 134 | | 124 | 125 |
| **36** | 115 | | 107 | | 116 | 122 |
| **37** | 122 | | 134 | | 123 | 115 |
| **38** | 105 | | 111 | | 113 | 113 |
| **39** | 103 | | 110 | | 114 | 103 |
| **40** | 105 | | 110 | | 93 | 90 |
| **41** | 97 | | 93 | | 104 | 84 |
| **42** | 88 | | 105 | | 76 | 114 |
| **43** | 104 | | 69 | | 93 | 89 |
| **44** | 87 | | 78 | | 79 | 74 |
| **45** | 77 | | 76 | | 76 | 66 |
| **46** | 72 | | 77 | | 61 | 54 |
| **47** | 75 | | 52 | | 57 | 68 |
| **48** | 62 | | 58 | | 68 | 69 |
| **49** | 57 | | 59 | | 60 | 57 |
| **50** | 43 | | 47 | | 50 | 42 |

| Intensity | Port #1 | Port #2 | Port #3 | Port #4 | Port #5 | Port #b |
|---|---|---|---|---|---|---|
| **51** | 37 | | 45 | | 35 | 38 |
| **52** | 36 | | 50 | | 34 | 36 |
| **53** | 41 | | 28 | | 38 | 38 |
| **54** | 18 | | 28 | | 26 | 10 |
| **55** | 23 | | 19 | | 17 | 27 |
| **56** | 19 | | 14 | | 20 | 24 |
| **57** | 17 | | 16 | | 17 | 21 |
| **58** | 7 | | 15 | | 8 | 6 |
| **59** | 8 | | 9 | | 6 | 9 |
| **60** | 9 | | 5 | | 2 | 6 |
| **61** | 3 | | 8 | | 2 | 7 |
| **62** | 2 | | 6 | | 4 | 8 |
| **63** | 6 | | 2 | | 3 | 2 |
| **64** | 3 | | 3 | | 1 | 3 |
| **65** | 4 | | 0 | | 1 | 3 |
| **66** | 3 | | 2 | | 0 | 3 |
| **67** | 2 | | 2 | | 0 | 0 |
| **68** | 0 | | 1 | | 2 | 2 |
| **69** | 0 | | 1 | | 0 | 0 |
| **70** | 0 | | 1 | | 0 | 0 |
| **71** | 0 | | 0 | | 0 | 0 |
| **72** | 0 | | 0 | | 0 | 0 |
| **73** | 0 | | 0 | | 0 | 0 |
| **74** | 0 | | 0 | | 1 | 0 |
| **75** | 0 | | 0 | | 0 | 0 |
| **76** | 0 | | 0 | | 0 | 0 |
| **77** | 0 | | 0 | | 0 | 0 |
| **78** | 0 | | 0 | | 0 | 0 |
| **79** | 0 | | 0 | | 0 | 0 |
| **80** | 0 | | 0 | | 0 | 0 |
| **81** | 0 | | 0 | | 0 | 0 |
| **82** | 0 | | 0 | | 0 | 0 |
| **83** | 0 | | 0 | | 0 | 0 |
| **84** | 0 | | 0 | | 0 | 0 |
| **85** | 0 | | 0 | | 0 | 0 |
| **86** | 0 | | 0 | | 0 | 0 |
| **87** | 0 | | 0 | | 0 | 0 |
| **88** | 0 | | 0 | | 0 | 0 |
| **89** | 0 | | 0 | | 0 | 0 |
| **90** | 0 | | 0 | | 0 | 0 |
| **91** | 0 | | 0 | | 0 | 0 |
| **92** | 0 | | 0 | | 0 | 0 |
| **93** | 0 | | 0 | | 0 | 0 |
| **94** | 0 | | 0 | | 0 | 0 |
| **95** | 0 | | 0 | | 0 | 0 |
| **96** | 0 | | 0 | | 0 | 0 |
| **97** | 0 | | 0 | | 0 | 0 |
| **98** | 0 | | 0 | | 0 | 0 |
| **99** | 0 | | 0 | | 0 | 0 |
| **100** | 0 | | 0 | | 0 | 0 |

**Phase II Full-Scale Experiment**
**Test #** 6 **Chamber #** 1
**Pre-ozonated 0.94 μm YG-Microspheres**

| Intensity | Port #1 | Port #2 | Port #3 | Port #4 | Port #5 | Port #b |
|---|---|---|---|---|---|---|
| **1** | 11 | 9 | | 17 | 14 | 21 |
| **2** | 48 | 84 | | 67 | 65 | 66 |
| **3** | 160 | 168 | | 159 | 124 | 138 |
| **4** | 162 | 170 | | 176 | 123 | 157 |
| **5** | 121 | 155 | | 128 | 68 | 124 |
| **6** | 77 | 103 | | 75 | 49 | 110 |
| **7** | 45 | 58 | | 36 | 24 | 86 |
| **8** | 14 | 34 | | 11 | 11 | 47 |
| **9** | 9 | 7 | | 14 | 9 | 22 |
| **10** | 6 | 12 | | 6 | 6 | 19 |
| **11** | 6 | 5 | | 4 | 3 | 12 |
| **12** | 2 | 4 | | 5 | 6 | 9 |
| **13** | 6 | 2 | | 2 | 5 | 6 |
| **14** | 0 | 5 | | 6 | 4 | 12 |
| **15** | 3 | 5 | | 1 | 9 | 9 |
| **16** | 1 | 2 | | 1 | 4 | 13 |
| **17** | 5 | 4 | | 10 | 5 | 19 |
| **18** | 3 | 7 | | 6 | 4 | 37 |
| **19** | 1 | 13 | | 4 | 1 | 26 |
| **20** | 6 | 5 | | 2 | 11 | 36 |
| **21** | 4 | 7 | | 3 | 10 | 40 |
| **22** | 4 | 10 | | 6 | 12 | 52 |
| **23** | 5 | 13 | | 15 | 19 | 54 |
| **24** | 4 | 16 | | 14 | 18 | 65 |
| **25** | 9 | 18 | | 15 | 19 | 104 |
| **26** | 5 | 28 | | 23 | 26 | 83 |
| **27** | 7 | 24 | | 29 | 24 | 85 |
| **28** | 13 | 31 | | 23 | 25 | 95 |
| **29** | 15 | 23 | | 37 | 43 | 101 |
| **30** | 11 | 40 | | 44 | 47 | 108 |
| **31** | 17 | 40 | | 41 | 43 | 121 |
| **32** | 25 | 41 | | 43 | 48 | 119 |
| **33** | 28 | 46 | | 63 | 53 | 98 |
| **34** | 32 | 60 | | 63 | 60 | 97 |
| **35** | 31 | 47 | | 84 | 56 | 124 |
| **36** | 30 | 65 | | 70 | 62 | 102 |
| **37** | 41 | 68 | | 70 | 81 | 112 |
| **38** | 42 | 56 | | 88 | 70 | 97 |
| **39** | 45 | 80 | | 105 | 88 | 89 |
| **40** | 54 | 72 | | 68 | 68 | 114 |
| **41** | 46 | 93 | | 87 | 81 | 86 |
| **42** | 40 | 59 | | 86 | 71 | 79 |
| **43** | 65 | 85 | | 90 | 111 | 108 |
| **44** | 53 | 86 | | 81 | 81 | 76 |
| **45** | 57 | 78 | | 84 | 83 | 75 |
| **46** | 77 | 90 | | 105 | 110 | 63 |
| **47** | 73 | 95 | | 100 | 96 | 72 |
| **48** | 76 | 89 | | 85 | 94 | 36 |
| **49** | 72 | 95 | | 111 | 106 | 37 |
| **50** | 59 | 97 | | 76 | 89 | 61 |

| Intensity | Port #1 | Port #2 | Port #3 | Port #4 | Port #5 | Port #b |
|---|---|---|---|---|---|---|
| **51** | 71 | 84 | | 80 | 108 | 39 |
| **52** | 79 | 96 | | 79 | 95 | 32 |
| **53** | 80 | 98 | | 84 | 99 | 24 |
| **54** | 67 | 96 | | 80 | 90 | 19 |
| **55** | 70 | 74 | | 76 | 72 | 27 |
| **56** | 71 | 61 | | 63 | 59 | 17 |
| **57** | 86 | 61 | | 73 | 75 | 12 |
| **58** | 89 | 69 | | 60 | 64 | 17 |
| **59** | 82 | 68 | | 69 | 85 | 13 |
| **60** | 67 | 57 | | 45 | 56 | 20 |
| **61** | 68 | 67 | | 64 | 63 | 7 |
| **62** | 82 | 37 | | 62 | 64 | 7 |
| **63** | 62 | 63 | | 39 | 45 | 7 |
| **64** | 76 | 31 | | 64 | 35 | 7 |
| **65** | 66 | 52 | | 33 | 36 | 4 |
| **66** | 73 | 41 | | 37 | 38 | 3 |
| **67** | 76 | 39 | | 29 | 43 | 4 |
| **68** | 49 | 32 | | 30 | 41 | 0 |
| **69** | 54 | 42 | | 21 | 38 | 0 |
| **70** | 61 | 24 | | 21 | 20 | 0 |
| **71** | 49 | 24 | | 22 | 19 | 1 |
| **72** | 46 | 24 | | 27 | 26 | 0 |
| **73** | 39 | 21 | | 17 | 21 | 0 |
| **74** | 54 | 12 | | 14 | 18 | 0 |
| **75** | 44 | 18 | | 7 | 15 | 0 |
| **76** | 34 | 11 | | 8 | 11 | 0 |
| **77** | 37 | 16 | | 5 | 9 | 0 |
| **78** | 30 | 5 | | 14 | 9 | 0 |
| **79** | 32 | 9 | | 4 | 11 | 0 |
| **80** | 34 | 13 | | 4 | 4 | 0 |
| **81** | 22 | 6 | | 5 | 2 | 0 |
| **82** | 21 | 8 | | 1 | 5 | 0 |
| **83** | 24 | 2 | | 0 | 3 | 0 |
| **84** | 20 | 2 | | 1 | 4 | 0 |
| **85** | 18 | 7 | | 4 | 2 | 0 |
| **86** | 14 | 4 | | 1 | 4 | 0 |
| **87** | 14 | 2 | | 3 | 2 | 0 |
| **88** | 13 | 4 | | 0 | 3 | 0 |
| **89** | 8 | 4 | | 0 | 0 | 0 |
| **90** | 9 | 3 | | 0 | 0 | 0 |
| **91** | 4 | 1 | | 1 | 1 | 0 |
| **92** | 7 | 0 | | 0 | 1 | 0 |
| **93** | 7 | 1 | | 0 | 0 | 0 |
| **94** | 7 | 1 | | 0 | 0 | 0 |
| **95** | 7 | 0 | | 0 | 0 | 0 |
| **96** | 6 | 0 | | 0 | 1 | 0 |
| **97** | 4 | 0 | | 0 | 0 | 0 |
| **98** | 4 | 1 | | 0 | 0 | 0 |
| **99** | 0 | 0 | | 0 | 0 | 0 |
| **100** | 1 | 1 | | 0 | 0 | 0 |

**Phase II Full-Scale Experiment**
**Test #** 6 **Chamber #** 2
**Pre-ozonated 0.94 $\mu$m YG-Microspheres**

| Intensity | Port #1 | Port #2 | Port #3 | Port #4 | Port #5 | Port #b |
|---|---|---|---|---|---|---|
| 1 | 11 | 12 | 5 | 15 | 3 | |
| 2 | 54 | 63 | 56 | 53 | 32 | |
| 3 | 120 | 112 | 132 | 107 | 48 | |
| 4 | 98 | 97 | 100 | 112 | 52 | |
| 5 | 64 | 77 | 97 | 66 | 36 | |
| 6 | 55 | 50 | 58 | 44 | 23 | |
| 7 | 32 | 26 | 46 | 31 | 17 | |
| 8 | 22 | 28 | 25 | 12 | 17 | |
| 9 | 17 | 12 | 16 | 7 | 9 | |
| 10 | 9 | 9 | 18 | 10 | 3 | |
| 11 | 4 | 13 | 15 | 7 | 9 | |
| 12 | 10 | 10 | 13 | 13 | 4 | |
| 13 | 10 | 20 | 15 | 14 | 5 | |
| 14 | 5 | 9 | 27 | 16 | 14 | |
| 15 | 17 | 22 | 29 | 16 | 13 | |
| 16 | 20 | 19 | 34 | 20 | 19 | |
| 17 | 29 | 27 | 49 | 25 | 14 | |
| 18 | 30 | 35 | 36 | 28 | 25 | |
| 19 | 39 | 41 | 60 | 40 | 31 | |
| 20 | 46 | 51 | 68 | 47 | 35 | |
| 21 | 50 | 57 | 79 | 48 | 48 | |
| 22 | 57 | 76 | 86 | 50 | 57 | |
| 23 | 66 | 82 | 107 | 73 | 54 | |
| 24 | 77 | 100 | 105 | 65 | 60 | |
| 25 | 79 | 104 | 114 | 79 | 65 | |
| 26 | 75 | 89 | 122 | 85 | 83 | |
| 27 | 114 | 96 | 129 | 72 | 86 | |
| 28 | 132 | 104 | 116 | 112 | 99 | |
| 29 | 106 | 126 | 147 | 91 | 108 | |
| 30 | 111 | 138 | 148 | 122 | 110 | |
| 31 | 116 | 134 | 126 | 119 | 112 | |
| 32 | 99 | 134 | 132 | 126 | 91 | |
| 33 | 111 | 109 | 138 | 121 | 115 | |
| 34 | 129 | 141 | 115 | 147 | 113 | |
| 35 | 112 | 122 | 119 | 105 | 122 | |
| 36 | 123 | 81 | 96 | 90 | 121 | |
| 37 | 107 | 110 | 107 | 131 | 115 | |
| 38 | 111 | 115 | 94 | 100 | 121 | |
| 39 | 83 | 84 | 79 | 98 | 92 | |
| 40 | 91 | 82 | 100 | 96 | 91 | |
| 41 | 104 | 100 | 77 | 78 | 102 | |
| 42 | 77 | 76 | 68 | 101 | 100 | |
| 43 | 92 | 71 | 57 | 67 | 86 | |
| 44 | 63 | 65 | 69 | 71 | 81 | |
| 45 | 72 | 75 | 70 | 67 | 60 | |
| 46 | 68 | 69 | 49 | 65 | 74 | |
| 47 | 55 | 72 | 33 | 58 | 63 | |
| 48 | 49 | 60 | 40 | 52 | 54 | |
| 49 | 61 | 37 | 35 | 63 | 54 | |
| 50 | 41 | 31 | 26 | 40 | 51 | |

| Intensity | Port #1 | Port #2 | Port #3 | Port #4 | Port #5 | Port #b |
|---|---|---|---|---|---|---|
| 51 | 39 | 29 | 33 | 55 | 41 | |
| 52 | 36 | 26 | 21 | 41 | 29 | |
| 53 | 35 | 21 | 10 | 33 | 42 | |
| 54 | 26 | 21 | 15 | 21 | 33 | |
| 55 | 15 | 11 | 8 | 22 | 31 | |
| 56 | 13 | 8 | 7 | 23 | 26 | |
| 57 | 9 | 13 | 5 | 13 | 27 | |
| 58 | 7 | 7 | 8 | 12 | 11 | |
| 59 | 7 | 4 | 2 | 17 | 15 | |
| 60 | 6 | 3 | 3 | 11 | 13 | |
| 61 | 9 | 5 | 5 | 8 | 12 | |
| 62 | 8 | 2 | 4 | 6 | 5 | |
| 63 | 4 | 3 | 1 | 8 | 8 | |
| 64 | 2 | 1 | 1 | 3 | 7 | |
| 65 | 2 | 1 | 1 | 3 | 3 | |
| 66 | 3 | 1 | 1 | 1 | 4 | |
| 67 | 0 | 0 | 0 | 3 | 6 | |
| 68 | 2 | 0 | 0 | 1 | 1 | |
| 69 | 1 | 1 | 3 | 2 | 3 | |
| 70 | 0 | 0 | 0 | 0 | 2 | |
| 71 | 1 | 0 | 0 | 0 | 0 | |
| 72 | 1 | 0 | 0 | 0 | 1 | |
| 73 | 0 | 0 | 0 | 0 | 1 | |
| 74 | 0 | 1 | 0 | 0 | 0 | |
| 75 | 0 | 0 | 0 | 0 | 0 | |
| 76 | 0 | 0 | 0 | 0 | 0 | |
| 77 | 0 | 0 | 0 | 0 | 0 | |
| 78 | 0 | 0 | 0 | 0 | 0 | |
| 79 | 0 | 0 | 0 | 0 | 0 | |
| 80 | 0 | 0 | 0 | 0 | 0 | |
| 81 | 0 | 0 | 0 | 0 | 0 | |
| 82 | 0 | 0 | 0 | 0 | 0 | |
| 83 | 0 | 0 | 0 | 0 | 0 | |
| 84 | 0 | 0 | 0 | 0 | 1 | |
| 85 | 0 | 0 | 0 | 0 | 0 | |
| 86 | 0 | 0 | 0 | 0 | 0 | |
| 87 | 0 | 0 | 0 | 0 | 0 | |
| 88 | 0 | 0 | 0 | 0 | 0 | |
| 89 | 0 | 0 | 0 | 0 | 0 | |
| 90 | 0 | 0 | 0 | 0 | 0 | |
| 91 | 0 | 0 | 0 | 0 | 0 | |
| 92 | 0 | 0 | 0 | 0 | 0 | |
| 93 | 0 | 0 | 0 | 0 | 0 | |
| 94 | 0 | 0 | 0 | 0 | 0 | |
| 95 | 0 | 0 | 0 | 0 | 0 | |
| 96 | 0 | 0 | 0 | 0 | 0 | |
| 97 | 0 | 0 | 0 | 0 | 0 | |
| 98 | 0 | 0 | 0 | 0 | 0 | |
| 99 | 0 | 0 | 0 | 0 | 0 | |
| 100 | 0 | 0 | 0 | 0 | 0 | |

**Phase II Full-Scale Experiment**
**Test #** 6 **Chamber #** 3
**Pre-ozonated 0.94 μm YG-Microspheres**

| Intensity | Port #1 | Port #2 | Port #3 | Port #4 | Port #5 | Port #b |
|---|---|---|---|---|---|---|
| 1 | 3 | 4 | 7 | 17 | 15 | 8 |
| 2 | 24 | 27 | 45 | 88 | 79 | 40 |
| 3 | 50 | 67 | 100 | 186 | 171 | 76 |
| 4 | 51 | 46 | 130 | 191 | 150 | 65 |
| 5 | 23 | 34 | 72 | 123 | 122 | 39 |
| 6 | 16 | 18 | 32 | 72 | 73 | 19 |
| 7 | 10 | 6 | 14 | 67 | 52 | 16 |
| 8 | 5 | 5 | 14 | 32 | 31 | 9 |
| 9 | 8 | 3 | 8 | 20 | 19 | 5 |
| 10 | 10 | 9 | 8 | 12 | 24 | 10 |
| 11 | 13 | 3 | 5 | 15 | 17 | 7 |
| 12 | 14 | 13 | 6 | 17 | 12 | 9 |
| 13 | 14 | 9 | 13 | 19 | 24 | 13 |
| 14 | 31 | 18 | 12 | 29 | 26 | 29 |
| 15 | 31 | 25 | 18 | 24 | 47 | 34 |
| 16 | 35 | 31 | 17 | 51 | 43 | 37 |
| 17 | 36 | 27 | 16 | 49 | 45 | 47 |
| 18 | 56 | 45 | 45 | 60 | 51 | 68 |
| 19 | 70 | 49 | 47 | 58 | 81 | 91 |
| 20 | 72 | 65 | 46 | 95 | 71 | 85 |
| 21 | 95 | 81 | 61 | 92 | 75 | 106 |
| 22 | 71 | 54 | 62 | 96 | 126 | 86 |
| 23 | 118 | 76 | 69 | 98 | 129 | 120 |
| 24 | 96 | 93 | 73 | 126 | 127 | 132 |
| 25 | 134 | 91 | 69 | 146 | 135 | 140 |
| 26 | 121 | 105 | 82 | 118 | 145 | 147 |
| 27 | 121 | 127 | 88 | 156 | 141 | 148 |
| 28 | 131 | 114 | 97 | 131 | 141 | 129 |
| 29 | 118 | 123 | 102 | 131 | 143 | 132 |
| 30 | 137 | 130 | 98 | 148 | 144 | 150 |
| 31 | 148 | 117 | 88 | 125 | 130 | 143 |
| 32 | 119 | 102 | 85 | 139 | 123 | 148 |
| 33 | 111 | 108 | 85 | 126 | 135 | 126 |
| 34 | 104 | 92 | 84 | 128 | 113 | 127 |
| 35 | 120 | 88 | 70 | 115 | 113 | 107 |
| 36 | 93 | 117 | 56 | 87 | 115 | 90 |
| 37 | 123 | 93 | 68 | 100 | 89 | 90 |
| 38 | 92 | 63 | 72 | 109 | 82 | 77 |
| 39 | 92 | 61 | 70 | 81 | 62 | 74 |
| 40 | 92 | 79 | 47 | 63 | 60 | 58 |
| 41 | 76 | 68 | 35 | 69 | 83 | 64 |
| 42 | 60 | 60 | 49 | 64 | 57 | 38 |
| 43 | 60 | 67 | 33 | 45 | 43 | 55 |
| 44 | 50 | 35 | 27 | 34 | 33 | 29 |
| 45 | 46 | 33 | 37 | 26 | 26 | 28 |
| 46 | 33 | 32 | 14 | 33 | 24 | 22 |
| 47 | 35 | 29 | 12 | 25 | 15 | 18 |
| 48 | 34 | 27 | 21 | 33 | 25 | 15 |
| 49 | 20 | 19 | 14 | 18 | 11 | 15 |
| 50 | 9 | 13 | 6 | 20 | 7 | 11 |

| Intensity | Port #1 | Port #2 | Port #3 | Port #4 | Port #5 | Port #b |
|---|---|---|---|---|---|---|
| 51 | 16 | 17 | 15 | 6 | 7 | 13 |
| 52 | 7 | 9 | 11 | 15 | 4 | 3 |
| 53 | 10 | 2 | 9 | 5 | 2 | 3 |
| 54 | 5 | 7 | 3 | 4 | 5 | 3 |
| 55 | 4 | 2 | 5 | 5 | 2 | 3 |
| 56 | 1 | 4 | 1 | 5 | 1 | 2 |
| 57 | 1 | 0 | 3 | 3 | 1 | 0 |
| 58 | 4 | 2 | 4 | 1 | 0 | 0 |
| 59 | 1 | 0 | 1 | 1 | 0 | 1 |
| 60 | 1 | 1 | 3 | 0 | 1 | 0 |
| 61 | 1 | 1 | 0 | 0 | 0 | 0 |
| 62 | 1 | 0 | 0 | 0 | 0 | 0 |
| 63 | 1 | 0 | 1 | 0 | 0 | 0 |
| 64 | 0 | 0 | 0 | 0 | 0 | 0 |
| 65 | 0 | 0 | 0 | 0 | 0 | 0 |
| 66 | 0 | 0 | 0 | 0 | 1 | 0 |
| 67 | 0 | 0 | 0 | 0 | 0 | 0 |
| 68 | 0 | 0 | 0 | 0 | 0 | 0 |
| 69 | 0 | 0 | 0 | 0 | 1 | 0 |
| 70 | 0 | 0 | 0 | 0 | 0 | 0 |
| 71 | 0 | 0 | 0 | 0 | 0 | 0 |
| 72 | 0 | 0 | 0 | 0 | 0 | 0 |
| 73 | 0 | 0 | 0 | 0 | 0 | 0 |
| 74 | 0 | 0 | 0 | 0 | 0 | 0 |
| 75 | 0 | 0 | 0 | 0 | 0 | 0 |
| 76 | 0 | 0 | 0 | 0 | 0 | 0 |
| 77 | 0 | 0 | 0 | 0 | 0 | 0 |
| 78 | 0 | 0 | 0 | 0 | 0 | 0 |
| 79 | 0 | 0 | 0 | 0 | 0 | 0 |
| 80 | 0 | 0 | 0 | 0 | 0 | 0 |
| 81 | 0 | 0 | 0 | 0 | 0 | 0 |
| 82 | 0 | 0 | 0 | 0 | 0 | 0 |
| 83 | 0 | 0 | 0 | 0 | 0 | 0 |
| 84 | 0 | 0 | 0 | 0 | 0 | 0 |
| 85 | 0 | 0 | 0 | 0 | 0 | 0 |
| 86 | 0 | 0 | 0 | 0 | 0 | 0 |
| 87 | 0 | 0 | 0 | 0 | 0 | 0 |
| 88 | 0 | 0 | 0 | 0 | 0 | 0 |
| 89 | 0 | 0 | 0 | 0 | 0 | 0 |
| 90 | 0 | 0 | 0 | 0 | 0 | 0 |
| 91 | 0 | 0 | 0 | 0 | 0 | 0 |
| 92 | 0 | 0 | 0 | 0 | 0 | 0 |
| 93 | 0 | 0 | 0 | 0 | 0 | 0 |
| 94 | 0 | 0 | 0 | 0 | 0 | 0 |
| 95 | 0 | 0 | 0 | 0 | 0 | 0 |
| 96 | 0 | 0 | 0 | 0 | 0 | 0 |
| 97 | 0 | 0 | 0 | 0 | 0 | 0 |
| 98 | 0 | 0 | 0 | 0 | 0 | 0 |
| 99 | 0 | 0 | 0 | 0 | 0 | 0 |
| 100 | 0 | 0 | 0 | 0 | 0 | 0 |

**Phase II Full-Scale Experiment**
**Test #** 6 **Chamber #** 4
**Pre-ozonated 0.94 $\mu$m YG-Microspheres**

| Intensity | Port #1 | Port #2 | Port #3 | Port #4 | Port #5 | Port #b |
|---|---|---|---|---|---|---|
| **1** | | | 13 | 0 | 19 | 3 |
| **2** | | | 84 | 5 | 95 | 43 |
| **3** | | | 140 | 11 | 146 | 74 |
| **4** | | | 151 | 7 | 119 | 54 |
| **5** | | | 110 | 6 | 110 | 54 |
| **6** | | | 45 | 7 | 65 | 37 |
| **7** | | | 43 | 2 | 28 | 36 |
| **8** | | | 18 | 4 | 15 | 18 |
| **9** | | | 7 | 3 | 11 | 15 |
| **10** | | | 17 | 9 | 18 | 15 |
| **11** | | | 23 | 10 | 13 | 19 |
| **12** | | | 28 | 11 | 16 | 23 |
| **13** | | | 37 | 14 | 27 | 26 |
| **14** | | | 36 | 30 | 35 | 22 |
| **15** | | | 47 | 29 | 35 | 47 |
| **16** | | | 62 | 33 | 47 | 55 |
| **17** | | | 79 | 54 | 69 | 59 |
| **18** | | | 99 | 55 | 76 | 86 |
| **19** | | | 98 | 79 | 102 | 87 |
| **20** | | | 130 | 85 | 113 | 101 |
| **21** | | | 111 | 93 | 107 | 114 |
| **22** | | | 134 | 115 | 144 | 121 |
| **23** | | | 127 | 127 | 143 | 167 |
| **24** | | | 134 | 124 | 138 | 139 |
| **25** | | | 167 | 147 | 143 | 161 |
| **26** | | | 161 | 139 | 145 | 157 |
| **27** | | | 151 | 172 | 140 | 148 |
| **28** | | | 167 | 138 | 172 | 157 |
| **29** | | | 147 | 138 | 177 | 149 |
| **30** | | | 124 | 132 | 158 | 148 |
| **31** | | | 125 | 138 | 150 | 128 |
| **32** | | | 138 | 148 | 139 | 133 |
| **33** | | | 120 | 131 | 99 | 122 |
| **34** | | | 82 | 119 | 118 | 104 |
| **35** | | | 92 | 93 | 109 | 86 |
| **36** | | | 69 | 84 | 100 | 80 |
| **37** | | | 78 | 93 | 88 | 81 |
| **38** | | | 72 | 76 | 76 | 89 |
| **39** | | | 54 | 76 | 50 | 55 |
| **40** | | | 42 | 67 | 61 | 53 |
| **41** | | | 57 | 54 | 45 | 39 |
| **42** | | | 35 | 50 | 44 | 41 |
| **43** | | | 33 | 44 | 29 | 30 |
| **44** | | | 24 | 40 | 26 | 30 |
| **45** | | | 14 | 27 | 27 | 15 |
| **46** | | | 20 | 26 | 23 | 12 |
| **47** | | | 14 | 26 | 18 | 16 |
| **48** | | | 6 | 13 | 11 | 9 |
| **49** | | | 7 | 21 | 9 | 5 |
| **50** | | | 6 | 14 | 4 | 9 |

| Intensity | Port #1 | Port #2 | Port #3 | Port #4 | Port #5 | Port #b |
|---|---|---|---|---|---|---|
| **51** | | | 3 | 12 | 7 | 4 |
| **52** | | | 8 | 4 | 5 | 1 |
| **53** | | | 1 | 4 | 1 | 2 |
| **54** | | | 0 | 1 | 2 | 0 |
| **55** | | | 0 | 0 | 0 | 1 |
| **56** | | | 0 | 2 | 1 | 1 |
| **57** | | | 0 | 1 | 0 | 1 |
| **58** | | | 1 | 2 | 0 | 0 |
| **59** | | | 0 | 0 | 0 | 2 |
| **60** | | | 0 | 0 | 0 | 0 |
| **61** | | | 0 | 0 | 0 | 0 |
| **62** | | | 0 | 0 | 0 | 0 |
| **63** | | | 1 | 0 | 0 | 0 |
| **64** | | | 0 | 0 | 0 | 1 |
| **65** | | | 0 | 0 | 0 | 0 |
| **66** | | | 0 | 0 | 0 | 0 |
| **67** | | | 0 | 0 | 0 | 0 |
| **68** | | | 0 | 0 | 0 | 0 |
| **69** | | | 0 | 0 | 0 | 0 |
| **70** | | | 1 | 0 | 0 | 0 |
| **71** | | | 0 | 0 | 0 | 0 |
| **72** | | | 1 | 0 | 0 | 0 |
| **73** | | | 0 | 0 | 0 | 0 |
| **74** | | | 0 | 1 | 0 | 0 |
| **75** | | | 0 | 0 | 0 | 0 |
| **76** | | | 0 | 0 | 0 | 0 |
| **77** | | | 0 | 0 | 0 | 0 |
| **78** | | | 0 | 0 | 0 | 0 |
| **79** | | | 0 | 0 | 0 | 0 |
| **80** | | | 0 | 0 | 0 | 0 |
| **81** | | | 0 | 0 | 0 | 0 |
| **82** | | | 0 | 0 | 0 | 0 |
| **83** | | | 0 | 0 | 0 | 0 |
| **84** | | | 0 | 0 | 0 | 0 |
| **85** | | | 0 | 0 | 0 | 0 |
| **86** | | | 0 | 0 | 0 | 0 |
| **87** | | | 0 | 0 | 0 | 0 |
| **88** | | | 0 | 0 | 0 | 0 |
| **89** | | | 0 | 0 | 0 | 0 |
| **90** | | | 0 | 0 | 0 | 0 |
| **91** | | | 0 | 0 | 0 | 0 |
| **92** | | | 0 | 0 | 0 | 0 |
| **93** | | | 0 | 0 | 0 | 0 |
| **94** | | | 0 | 0 | 0 | 0 |
| **95** | | | 0 | 0 | 0 | 0 |
| **96** | | | 0 | 0 | 0 | 0 |
| **97** | | | 0 | 0 | 0 | 0 |
| **98** | | | 0 | 0 | 0 | 0 |
| **99** | | | 0 | 0 | 0 | 0 |
| **100** | | | 0 | 0 | 0 | 0 |

**Phase II Full-Scale Experiment**
**Test #** 6 **Chamber #** 5
**Pre-ozonated 0.94 μm YG-Microspheres**

| Intensity | Port #1 | Port #2 | Port #3 | Port #4 | Port #5 | Port #b |
|---|---|---|---|---|---|---|
| **1** | 2 | 4 | 6 | 0 | | 5 |
| **2** | 20 | 35 | 54 | 24 | | 51 |
| **3** | 33 | 62 | 86 | 55 | | 105 |
| **4** | 38 | 67 | 86 | 66 | | 97 |
| **5** | 26 | 55 | 56 | 52 | | 42 |
| **6** | 16 | 49 | 37 | 48 | | 21 |
| **7** | 9 | 22 | 15 | 24 | | 12 |
| **8** | 2 | 11 | 6 | 18 | | 10 |
| **9** | 5 | 3 | 8 | 7 | | 10 |
| **10** | 4 | 15 | 6 | 7 | | 8 |
| **11** | 4 | 7 | 14 | 10 | | 15 |
| **12** | 8 | 21 | 11 | 21 | | 18 |
| **13** | 8 | 26 | 15 | 13 | | 17 |
| **14** | 17 | 30 | 31 | 22 | | 23 |
| **15** | 14 | 36 | 40 | 36 | | 40 |
| **16** | 25 | 55 | 40 | 61 | | 45 |
| **17** | 21 | 56 | 64 | 50 | | 52 |
| **18** | 27 | 67 | 63 | 67 | | 56 |
| **19** | 35 | 79 | 98 | 85 | | 79 |
| **20** | 26 | 85 | 104 | 81 | | 95 |
| **21** | 43 | 90 | 124 | 110 | | 98 |
| **22** | 41 | 110 | 123 | 105 | | 117 |
| **23** | 38 | 109 | 120 | 127 | | 122 |
| **24** | 51 | 134 | 136 | 104 | | 121 |
| **25** | 56 | 135 | 138 | 136 | | 127 |
| **26** | 52 | 125 | 125 | 133 | | 139 |
| **27** | 53 | 124 | 119 | 129 | | 139 |
| **28** | 50 | 130 | 103 | 131 | | 117 |
| **29** | 46 | 106 | 121 | 122 | | 148 |
| **30** | 49 | 119 | 109 | 96 | | 119 |
| **31** | 53 | 115 | 99 | 106 | | 91 |
| **32** | 36 | 110 | 123 | 103 | | 106 |
| **33** | 41 | 98 | 97 | 107 | | 98 |
| **34** | 33 | 90 | 90 | 74 | | 97 |
| **35** | 38 | 78 | 75 | 68 | | 84 |
| **36** | 22 | 87 | 56 | 80 | | 65 |
| **37** | 23 | 68 | 49 | 50 | | 66 |
| **38** | 22 | 45 | 49 | 49 | | 69 |
| **39** | 18 | 48 | 47 | 48 | | 39 |
| **40** | 15 | 44 | 39 | 48 | | 34 |
| **41** | 18 | 29 | 38 | 31 | | 25 |
| **42** | 10 | 16 | 27 | 32 | | 37 |
| **43** | 9 | 20 | 19 | 33 | | 21 |
| **44** | 6 | 17 | 22 | 15 | | 24 |
| **45** | 6 | 15 | 14 | 24 | | 22 |
| **46** | 4 | 13 | 8 | 14 | | 15 |
| **47** | 5 | 10 | 5 | 2 | | 9 |
| **48** | 1 | 7 | 6 | 6 | | 9 |
| **49** | 5 | 4 | 3 | 5 | | 7 |
| **50** | 0 | 3 | 2 | 7 | | 4 |

| Intensity | Port #1 | Port #2 | Port #3 | Port #4 | Port #5 | Port #b |
|---|---|---|---|---|---|---|
| **51** | 1 | 2 | 6 | 3 | | 1 |
| **52** | 2 | 4 | 3 | 0 | | 4 |
| **53** | 1 | 0 | 3 | 1 | | 1 |
| **54** | 0 | 1 | 0 | 3 | | 0 |
| **55** | 0 | 1 | 0 | 0 | | 0 |
| **56** | 0 | 1 | 0 | 1 | | 1 |
| **57** | 0 | 1 | 0 | 0 | | 1 |
| **58** | 0 | 0 | 0 | 0 | | 0 |
| **59** | 0 | 0 | 1 | 0 | | 0 |
| **60** | 0 | 0 | 0 | 0 | | 0 |
| **61** | 0 | 0 | 0 | 0 | | 0 |
| **62** | 0 | 0 | 0 | 0 | | 0 |
| **63** | 0 | 0 | 2 | 0 | | 0 |
| **64** | 0 | 0 | 0 | 0 | | 0 |
| **65** | 0 | 0 | 0 | 0 | | 0 |
| **66** | 0 | 0 | 0 | 0 | | 1 |
| **67** | 0 | 0 | 0 | 0 | | 0 |
| **68** | 0 | 0 | 0 | 0 | | 0 |
| **69** | 0 | 0 | 0 | 0 | | 0 |
| **70** | 0 | 0 | 0 | 0 | | 0 |
| **71** | 0 | 0 | 0 | 0 | | 0 |
| **72** | 0 | 0 | 0 | 0 | | 0 |
| **73** | 0 | 0 | 0 | 0 | | 0 |
| **74** | 0 | 0 | 0 | 0 | | 0 |
| **75** | 0 | 0 | 0 | 0 | | 0 |
| **76** | 0 | 0 | 0 | 0 | | 0 |
| **77** | 0 | 0 | 0 | 0 | | 0 |
| **78** | 0 | 0 | 0 | 0 | | 0 |
| **79** | 0 | 0 | 0 | 0 | | 0 |
| **80** | 0 | 0 | 0 | 0 | | 0 |
| **81** | 0 | 0 | 0 | 0 | | 0 |
| **82** | 0 | 0 | 0 | 0 | | 0 |
| **83** | 0 | 0 | 0 | 0 | | 0 |
| **84** | 0 | 0 | 0 | 0 | | 0 |
| **85** | 0 | 0 | 0 | 0 | | 0 |
| **86** | 0 | 0 | 0 | 0 | | 0 |
| **87** | 0 | 0 | 0 | 0 | | 0 |
| **88** | 0 | 0 | 0 | 0 | | 0 |
| **89** | 0 | 0 | 0 | 0 | | 0 |
| **90** | 0 | 0 | 0 | 0 | | 0 |
| **91** | 0 | 0 | 0 | 0 | | 0 |
| **92** | 0 | 0 | 0 | 0 | | 0 |
| **93** | 0 | 0 | 0 | 0 | | 0 |
| **94** | 0 | 0 | 0 | 0 | | 0 |
| **95** | 0 | 0 | 0 | 0 | | 0 |
| **96** | 0 | 0 | 0 | 0 | | 0 |
| **97** | 0 | 0 | 0 | 0 | | 0 |
| **98** | 0 | 0 | 0 | 0 | | 0 |
| **99** | 0 | 0 | 0 | 0 | | 0 |
| **100** | 0 | 0 | 0 | 0 | | 0 |

## REFERENCES

American National Standards Institute, NSF International (2000). *Drinking Water Treatment Chemicals - Health Effects.* ANSI/NSF 60-2000. Ann Arbor, MI.

American Public Health Association (APHA), American Water Works Association (AWWA), Water Environment Federation (WEF). (1998). *Standard Methods for the Examination of Water and Wastewater*, American Public Health Association, Washington, D.C.

Chiou, C.-F., Mariñas B. J., Adams J. Q. (1995). "Modified Indigo Method for Gaseous and Aqueous Ozone Analysis", *Ozone Science & Engineering*, 17:329-344.

Chiou, C-F., Torres-Lugo, M., Mariñas, B. J., Adams, J. Q. (1997). "Non-biological Surrogate Indicators for Assessing Ozone Disinfection", *Journal American Water Works Association*, 89(8):54-66.

Clark, R. M., Sivagenesan, M., Rice, E. W., Chen, J. (2002). "Development of a *Ct* Equation for the inactivation of *Cryptosporidium* oocysts with ozone", *Water Research*, 36(12):3141-3149.

Corona-Vasquez, B., Samuelson, A., Rennecker, J. L., Mariñas, B. J. (2002) "Inactivation of *Cryptosporidium parvum* oocysts with ozone and free chlorine", *Water Research*, 36:4053-4063.

Driedger, A. M., Rennecker, J. L., Mariñas, B. J. (2000). "Sequential Inactivation of *Cryptosporidium parvum* Oocysts with Ozone and Free Chlorine", *Water Research*, 34(14):3591-3597.

Echigo, S., Minear, R. A., Yamada, H. (1999). "Ultra Low Bromate Detection in Drinking Water", *Extended Abstracts of 217$^{th}$ ACS National Meeting, Division of Environmental Chemistry, March, Anaheim, CA* 39(1):249-251.

Hirata, T., Shimura, A., Morita, S., Suzuki, M., Motoyama, N., Hoshikawa, H., Moniwa, T., Kaneko, M. (2001) "The effect of temperature on the efficacy of ozonation for inactivating *Cryptosporidium parvum* oocysts" *Water Science and Technology*, 43(12):163-166.

Kim, J.-H. (2002). Integrated Optimization of Bromate Formation and *Cryptosporidium parvum* Oocyst Control in Ozone Bubble-Diffuser Contactors, *Ph. D. Dissertation*, University of Illinois at Urbana-Champaign.

Kim, J.-H., Rennecker, J.L., Tomiak, R.B., Mariñas, B.J., Miltner, R.J., Owens, J.H. (2002a). "Inactivation of *Cryptosporidium* Oocysts in a Pilot-scale Bubble-diffuser Contactor – II.: Model Validation and Application", *ASCE Journal of Environmental Engineering*, 128(6):522-532.

Kim, J.-H.; Tomiak, R.B.; Mariñas, B.J. (2002b). "Inactivation of *Cryptosporidium* Oocysts in a Pilot-scale Bubble-diffuser Contactor – I.: Model Development", *ASCE Journal of Environmental Engineering*, 128(6):514-521.

Kim, J.-H., von Gunten, U., Mariñas, B.J. (2001). "Modeling the Effect of Water Quality on Bromate Formation and *Cryptosporidium parvum* Oocyst Inactivation during Ozone Treatment", *Proceedings of the 2001 American Water Works Association Annual Conference*, Washington, DC, June 17-21, 2001.

Kramer, M. H., Herwaldt, B. L., Craun, G. F. (1996). "Waterborne Disease: 1993 and 1994", *Journal American Water Works Association*, 88(3):66-80.

Li, H., Finch, G. R., Smith, D. W., Belosevic, M. (2000). "Chemical Inactivation of *Cryptosporidium* in Water Treatment", Denver, CO: AWWA Research Foundation.

Mariñas B. J., Liang, S., and Aieta, M. E. (1993). "Modeling hydrodynamics and ozone residual distribution in a pilot-scale ozone bubble-diffuser contactor." *Journal American Water Works Association*, 85(3): 90-99.

Mariñas, B. J., Rennecker, J. L., Teefy, S., Rice, E. W. (1999). "Assessing Ozone Disinfection with Nonbiological Surrogates", *Journal American Water Works Association*, 91(9):79-89.

Oppenheimer, J. A., Aieta, E. M., Trusell, R. R., Jacangelo, J. G., Najm, N. (2000) "Evaluation of *Cryptosporidium* Inactivation in Natural Waters" Denver, CO: AWWA Research Foundation.

Owens, J. H., Miltner, R. J., Rice, E. W., Johnson, C. H., Dahling, D. R., Schaefer, F. W., Shukairy, H., M. (2000). "Pilot-scale ozone Inactivation of *Cryptosporidium* and other Microorganisms in Natural Water", *Ozone Science & Engineering*, 22:501-517.

Press, W. H., Teukolsky, S.A., Vetterling, W.T., Flannery, B.P. (1992) Numerical Recipes in C., Cambridge University Press, 2nd ed.

Rennecker, J. L., Mariñas B. J., Rice E. W., Owens J. H. (1999). "Inactivation of *Cryptosporidium parvum* Oocysts with Ozone", *Water Research*, 33(9):2481-2488.
Shapiro, H. M. (1995). *Practical Flow Cytometry*, 3rd ed., Wiley-Liss, Inc., New York, NY.

Rennecker, J. L., Driedger, A. M., Rubin, S. A., Mariñas, B. J. (2000). "Synergy in Sequential Inactivation of *Cryptosporidium parvum* with Ozone/Free Chlorine and Ozone/Monochloramine", *Water Research*, 34(17):4121-4130.

Rennecker, J. L., Kim, J-H., Corona-Vasquez, B., Mariñas, B. J. (2001). "Role of Disinfectant Concentration and pH in the Inactivation Kinetics of *Cryptosporidium parvum* Oocysts with Ozone and Monochloramine" *Environmental Science & Technology,* 13(7):2752-2757.

Scharfenaker, Mark A. (2002). "Draft LT2ESWTR out of the box" *Journal American Water Works Association,* 94(2):24037.

Shapiro, H. M. (1995) *Practical Flow Cytometry*, 3rd ed., Wiley-Liss, Inc., New York, NY.

Solo-Gabriele, H., Neumeister, S. (1996). "U.S. Outbreaks of Cryptosporidiosis", *Journal American Water Works Association*, 88(9):76-86.

United States Environmental Protection Agency. (1991). "National Primary Drinking Water Regulations: Disinfection, Turbidity, *Gardia lamblia*, Viruses, Legionella, and Heterotrophic Bacteria; Final Rule", *Federal Register*, 54(124):27485-27541.

United States Environmental Protection Agency. (1998a). "National Primary Drinking Water Regulations: Disinfectants and Disinfection By-products; Final Rule", *Federal Register*, 63(241):69389-69476.

United States Environmental Protection Agency. (1998b). "Water Contaminant Candidate List; Notice", *Federal Register*,63(40):10273-10287.

Weinberg, H. S., Yamada, H. (1998). "Post-Ion-Chromatography Derivatization for the Determination of Oxyhalides at sub-PPB Levels in Drinking Water", *Analytical Chemistry,* 70(1):1-6.

## ABBREVIATIONS

| | |
|---|---|
| ACWD | Alameda County Water District, Fremont, California |
| ANSI | American National Standards Institute |
| APHA | American Public Health Association |
| AWWA | American Water Works Association |
| AwwaRF | Awwa Research Foundation |
| $^{o}$C | degrees Celsius |
| $C$ | dissolved ozone concentration |
| CA DHS | State of California Department of Health Services |
| CCL | contaminant candidate list |
| cm | centimeter |
| CSTR | continuous-stirred tank reactor |
| $CT$ | product of disinfection concentration and contact time |
| $CT_{lag}$ | $CT$ required to overcome the initial time lag |
| $d$ | dispersion number |
| $D$ | fast demand concentration |
| D/DBP | disinfectants and disinfection by-products |
| DBP | disinfection by-products |
| EDA | ethylene diamine |
| $f_i$ | overall rate function corresponding to the relevant chemical reactions |
| FITC | fluorescein isothiocyanate |
| ft | foot |
| FSC | forward scattering |
| g/L | gram per liter |
| g/ml | gram per milliliter |
| HRT | hydraulic residence time |
| $I$ | average fluorescence intensity |
| $k_D$ | second-order rate constant for the reaction between ozone and fast demand |
| kJ/mol | kilojoule per mole |
| $k_N$ | pseudo first-order inactivation rate constant |
| $k_S$ | first-order rate constant for ozone decay |
| L/mg×m | liter per milligram per meter |
| LT2ESWTR | Long-Term 2 Enhanced Surface Water Treatment Rule |

| | |
|---|---|
| m | meter |
| M | molar |
| MCL | maximum contaminant level |
| MCLG | maximum contaminant level goal |
| MGD | mega gallon per day |
| $M^{-1}cm^{-1}$ | per molar per centimeter |
| mg×min/L | milligram minute per liter |
| mg/L | milligram per liter |
| mg-$O_3$/L | milligram ozone per liter |
| min | minute |
| mL | milliliter |
| mL/min | milliliter per minute |
| M/mg×min | molar per milligram per minute |
| mW | milliwatt |
| μg/L | microgram per liter |
| μm | micrometer |
| | |
| nm | nanometer |
| $N/N_o$ | survival ratio |
| $N_l/N_o$ | lag-phase factor |
| NSF | National Sanitation Foundation |
| NTU | nephelometric turbidity unit |
| | |
| PAC | Project Advisory Committee |
| PFR | plug-flow reactor |
| | |
| $s^{-1}$ | per second |
| SCFM | standardized cubic feet per minute |
| SEM | scanning electron microscopy |
| SSC | side scattering |
| SWTR | Surface Water Treatment Rule |
| | |
| t | time |
| $T$ | temperature in degree Celsius |
| $t_{10}$ | time for 10 percent of total mass of tracer to exit the reactor |
| TOC | total organic carbon |
| θ | hydraulic residence time in the chamber |
| | |
| USEPA | United States Environmental Protection Agency |
| | |
| WEF | Water Environment Federation |
| | |
| $z$ | normalized downward distance from the water surface |

IWA Publishing's authorised EU representative for General Product Safety Regulations is Diane D'Arras, 15 rue Duret, 75116 Paris, France, e-mail: safety@iwap.co.uk.

Printed and bound by CPI Group (UK) Ltd, Croydon, CR0 4YY
17/09/2026
02218798-0013